Introduction to the Replica Theory of Disordered Statistical Systems

This book provides a detailed introduction to the statistical mechanics of classical spin systems with quenched disorder. The first part of the book describes the physics of spin-glass states using results obtained within the framework of the mean field theory of spin glasses. The technique of replica symmetry breaking is explained in detail, along with a discussion of the underlying physics. The second part is devoted to the theory of critical phenomena in the presence of weak quenched disorder. Here one can find a systematic derivation of the traditional renormalization group theory, which is then used to obtain a new 'random' critical regime in disordered vector ferromagnets and in the two-dimensional Ising model. The third part of the book describes other types of disordered systems, with relation to new results at the frontiers of modern research. This book is suitable for graduate students and researchers in the field of statistical mechanics of disordered systems. It will also be useful as a supplementary text for graduate level courses in statistical mechanics.

VIKTOR DOTSENKO is a professor at the Laboratoire de Physique Théoretique des Liquides, Université Paris VI, and has been a researcher at the Landau Institute for Theoretical Physics in Moscow since 1983. He has lectured at universities all over the world and from 1990 to 1992 was a visiting professor at the universities of Rome, La Sapienza and Tor Vergata, and then at the Ecole Normal Supérieure, Paris. He has published widely in the area of statistical mechanics and disordered spin systems.

Collection Aléa–Saclay: Monographs and texts in statistical physics

C. Godrèche (ed.): Solids Far from Equilibrium
P. Peretto: An Introduction to the Modeling of Neural Networks
C. Godrèche and P. Manneville (eds.): Hydrodynamics and Nonlinear Instabilities
A. Pimpinelli and J. Villain: Physics of Crystal Growth
D. H. Rothman and S. Zaleski: Lattice-Gas Cellular Automata
B. Chopard and M. Droz: Cellular Automata Modeling of Physical Systems
J. Marro and R. Dickman: Nonequilibrium Phase Transitions in Lattice Models
J.-F. Sadoc and R. Mosseri: Geometrical Frustration
V. Dotsenko: Introduction to the Replica Theory of Disordered Statistical Systems

Introduction to the Replica Theory of Disordered Statistical Systems

Viktor Dotsenko
Université Paris VI
and
Landau Institute for Theoretical Physics, Moscow

CAMBRIDGE UNIVERSITY PRESS
Cambridge, New York, Melbourne, Madrid, Cape Town, Singapore, São Paulo

Cambridge University Press
The Edinburgh Building, Cambridge CB2 2RU, UK

Published in the United States of America by Cambridge University Press, New York

www.cambridge.org
Information on this title: www.cambridge.org/9780521773409

First published 2001
This digitally printed first paperback version 2005

A catalogue record for this publication is available from the British Library

ISBN-13 978-0-521-77340-9 hardback
ISBN-10 0-521-77340-7 hardback

ISBN-13 978-0-521-02125-8 paperback
ISBN-10 0-521-02125-1 paperback

Contents

Preface

This book is devoted to the special area of statistical mechanics that deals with the classical spin systems with quenched disorder. It is assumed to be of a pedagogical character, and it aims to help the reader to get into the subject starting from fundamentals. The book is supposed to be selfcontained (the reader is not required to go through all the references to understand something), being understandable for any student having basic knowledge of theoretical physics and statistical mechanics. Nevertheless, because this is only an *introduction* to the wide scope of statistical mechanics of disordered systems, in some cases to get to know more details about a particular topic the reader is advised to refer to the existing literature. Although throughout the book I have tried to present all the unavoidable calculations such that they would look as transparent as possible and have given everywhere (where it is at all possible) physical interpretations of what is going on, in many cases certain personal efforts and/or use of imagination are still required.

The **first part** of the book is devoted to the physics of spin-glass systems, where the quenched disorder is the dominant factor. The emphasis is made on a general qualitative description of the physical phenomena, being mostly based on the results obtained in the framework of the mean-field theory of spin-glasses with long-range interactions [1]. First, the general problems of the spin-glass state are discussed at the qualitative level [2]. In Chapters 3–5 the 'magic' of the replica symmetry breaking (RSB) scheme is explained in detail, and the physics behind it is discussed. This part also contains the detailed derivation of the ultrametric structure of the space of spin-glass states as well as its scaling properties (the reader interested in more details about ultrametricity may refer to the review [3]). Chapter 6 is devoted to the review of particular experimental works on real spin-glass materials, which on a qualitative level confirm the basic theoretical predictions.

It should be stressed that in this book I have no intentions of giving any kind of review of the very many different problems that exist in the scope of spin glasses. For that the reader should refer to the many excellent books and reviews that exist [1, 4, 5]. My intentions are much more modest. Here I would mostly like to present in a systematic way the *methods* and *physical ideas* of the replica theory, and in particular the powerful technique of the replica symmetry breaking. On the other hand (keeping in mind the pedagogical character of this book), I would also like to demonstrate how this *general* method works when applied to various (which often look quite different) *particular* problems of statistical mechanics, such as critical phenomena, directed polymers, Griffiths singularities, etc., which are considered in subsequent parts of the book. For those reasons I *do not* discuss here the exciting current debates about the validity of the considered mean-field approach for finite-dimensional spin glasses [5, 6], as well as the alternative point of view on the spin-glass state, which is called the droplet model [7]. For the same reasons I do not include consideration of the very important and presently actively developing field of *dynamical* behavior of spin glasses [8, 9]. At the moment these very interesting problems are still too far from being resolved.

The **second part** of the book is mainly devoted to the theory of the critical phenomena at the phase transitions of the second order in the presence of weak quenched disorder. The theory of the critical phenomena deals with macroscopic statistical systems in the close vicinity of the phase-transition point where spontaneous symmetry breaking takes place, and the situation is characterized by large-scale fluctuations. According to the traditional scaling theory the large-scale fluctuations are characterized by a certain dominant scale, the correlation length, R_c, which grows as the critical point is approached, where it becomes infinite. The large-scale fluctuations lead to singularities in the macroscopic characteristics of the system as a whole. These singularities are the main subject of the theory. Chapter 7 is devoted to the systematic consideration of the traditional renormalization group (RG) theory of the critical phenomena, including ϵ-expansion. The reader familar with this technique (which is well explained in many books, see e.g. [10]) may easily skip this chapter.

Originally, many years ago, it was generally believed that quenched disorder could either completely destroy the long-range fluctuations, such that the singularities of the thermodynamical functions become smoothed out, or it could produce only a shift of the critical point but could not affect the critical behavior itself. Later it was realized that an intermediate situation is also possible, in which a new critical behavior, with new universal critical exponents, is established sufficiently close to the phase-transition point. In terms of the RG approach the standard procedure for obtaining a new universal 'random' critical regime for the vector ferromagnetic spin systems is considered in Chapter 8.

However, according to recent developments in this field, the effects of the quenched disorder on the critical behavior could appear to be more complicated, and in certain cases a completely new type of critical phenomenon of the spin-glass nature could be established in the close vicinity of the critical point. In Chapter 9 the RG theory for the vector ferromagnetic spin systems is generalized to take into account these non-perturbative spin-glass-type phenomena. It is demonstrated that whenever the disorder is relevant for the critical behavior there exist no stable fixed points, and the RG flows lead to the so-called strong coupling regime at a finite spatial scale. The physical consequences of the obtained RG solutions are discussed.

In Chapter 10 we consider the critical properties of the two-dimensional disordered Ising model. It is well known that in the critical region the two-dimensional Ising model can be reduced to the free-fermion theory [11]. Here this reduction will be demonstrated in very simple terms by means of the Grassman variables technique (for detailed treatment of this new mathematics see [12–14]). The resulting continuum theory, to which the exact lattice-disordered model is equivalent in the critical region, appears to be simple enough, and its specific heat critical behavior can be found exactly. Here we also consider the results of the recent numerical simulations of the two-dimensional disordered Ising model, and describe the general structure of its phase diagram.

In the **third part** of the book some other types of disordered system are described. Here the reader is taken to the frontier of modern research, because this part is mostly devoted to problems which are not yet solved.

In Chapter 11 the Ising spin systems with quenched random fields are considered. This type of disorder is essentially different from that with random interactions, because external magnetic fields break the symmetry with respect to the change of the signs of spins. Despite extensive theoretical and experimental efforts during the past 20 years or so (for reviews see e.g. [5, 15]) very little is understood about even the basic thermodynamic properties of the random field Ising model (in particular the nature of the phase transition in this model is still a mystery). Here it is demonstrated why the statistical systems of this type exhibit many qualitatively different properties compared with those considered before.

In a wide variety of physical systems one is interested in the behavior of a fluctuating linear object (with finite line tension) interacting with a quenched random potential. This class of problem is traditionally discussed in terms of a directed polymer in random media, and it has been much studied during recent years [16]. Quite naturally the best understanding has been achieved for the simplest one-dimensional case when the displacements of a directed polymer can occur only in one direction. Chapter 12 is devoted to the derivation of the scaling properties of the one-dimensional directed polymers in random potentials with finite correlation radius. Here it is demonstrated that in the low-temperature limit the solution of the problem can be described in terms of the effective one-step replica symmetry breaking ansatz.

In Chapter 13, I present a new method to study statistical systems with quenched disorder in the low-temperature limit, the so-called vector replica symmetry breaking ansatz. At first sight in the low-temperature limit the situation must be simplified because the partition function could be analyzed in terms of the saddle-point approximation. However, it is easy to see that generically, even at zero temperature, there still exist sample-to-sample fluctuations. It appears that in a large class of strongly disordered systems, it is necessary to include saddle points of the Hamiltonian that break the replica symmetry in a *vector* sector, as opposed to the usual matrix sector breaking of spin-glass mean-field theory. First, I demonstrate how this new method works on the examples of some elementary problems which can be solved directly. As for its further application to more difficult problems, the zero-temperature fluctuations of a particle in a random potential are studied, and the scaling exponents of the directed polymers in random media with long-range correlations are rederived using this new method.

In the last section of this chapter I consider the problem of the existence of non-analytic (Griffiths-like) contributions to the free energy of a weakly disordered Ising ferromagnet (in a zero external magnetic field) from the point of view of the replica theory. Here it is demonstrated that in the paramagnetic phase (away from the critical point) such contributions appear as a result of non-linear localized (instanton-like) solutions of the mean-field saddle-point equations that are characterized by the vector type of the replica symmetry breaking.

1
Introduction

1.1 General principles of the statistical mechanics

In the most simple terms the basic statements of the statistical mechanics can be introduced in the following way. Let the microscopic state of a *macroscopic* system having many degrees of freedom be described by the configurations of N variables $\{s_i\}$, $(i = 1, 2, \ldots, N)$. The basic quantity characterizing the microscopic states is called the *energy*, H, and it is defined as a function of all the microscopic variables $\{s_i\}$:

$$H = H(s_1, s_2, \ldots, s_N) \equiv H[\mathbf{s}]$$

The microscopic dynamic behavior of the system is defined by some dynamic differential equations such that, in general, the energy of the system tends to a minimum. Besides, it is assumed that no observable system can be perfectly isolated from the surrounding world, and the effect of the interaction with the surroundings (the thermal bath) is believed to produce the so-called *thermal noise* in the exact dynamical equations. The thermal (white) noise acts as random and uncorrelated fluctuations, which produce the randomization and the mixing of the exact dynamical trajectories of the system.

Let $A[\mathbf{s}]$ be some observable quantity. The quantities, which are of interest in statistical mechanics, are the *averaged* values of the observables. In other words, instead of studying the exact evolution in time of the value $A[\mathbf{s}(t)]$, one introduces the averaged quantity:

$$\langle A \rangle = \lim_{t \to \infty} \frac{1}{t} \int_0^t dt' A[\mathbf{s}(t')] \qquad (1.1)$$

which could be formally obtained after the observation during an infinite time period.

The fundamental hypothesis of the equilibrium statistical mechanics lies in the following. It is believed that, owing to the mixing of the dynamic trajectories, after an infinitely long observation time, the system in general 'visits' its different microscopic states many times, and therefore the averaged quantity in Eq. (1.1) could be obtained by averaging over the *ensemble* of the states instead of that over the time:

$$\langle A \rangle = \int ds_1 ds_2 \ldots ds_N A[\mathbf{s}] P(s_1, s_2, \ldots, s_N) \tag{1.2}$$

Here $P[\mathbf{s}]$ is the *probability distribution function* of the microscopic states of the system. In other words, it is believed that because of the mixing of the dynamic trajectories, instead of solving the exact dynamics, the system could be statistically described in terms of the probabilities of its microscopic states given by the function $P[\mathbf{s}]$. The probability distribution function, whatever it is, must be normalized:

$$\int ds_1 ds_2 \ldots ds_N P(s_1, s_2, \ldots, s_N) = 1 \tag{1.3}$$

The fundamental quantity of the statistical mechanics that characterizes the probability distribution itself is called the entropy. It is defined as the average of the logarithm of the distribution function:

$$S = -\langle \log(P[\mathbf{s}]) \rangle \equiv -\int ds_1 ds_2 \ldots ds_N P[\mathbf{s}] \log(P[\mathbf{s}]) \tag{1.4}$$

In general, the value of the entropy could tell to what extent the state of the system is 'ordered'. Consider a simple illustrative example. Let the (discrete) microscopic states of the system be labeled by an index α, and let us assume that the probability distribution is such that only L (among all) states have non-zero and equal probability. Then, owing to the normalization (1.3), the probability of any of these L states must be equal to $1/L$. According to the definition of the entropy, one gets:

$$S = -\sum_{\alpha}^{L} P_\alpha \log P_\alpha = \log L$$

Therefore, the broader the distribution (i.e. the larger L) is, the larger the value of the entropy is. On the other hand, the more concentrated the distribution function is, the smaller the value of the entropy is. In the extreme case, when there is only one microscopic state occupied by the system, the entropy is equal to zero. In general, the value of $\exp(S)$ could be interpreted as the averaged number of the states occupied by the system with finite probability.

Now let us consider what the general form of the probability distribution function must be. According to the basic hypothesis, the average value of the energy of the system is:

$$E \equiv \langle H \rangle = \sum_{\alpha} P_{\alpha} H_{\alpha} \tag{1.5}$$

The interaction of the system with the thermal bath produces the following fundamental effects. First, the averaged value of its energy in the thermal equilibrium is conserved. Second, for some reason Nature is constructed in such a way that irrespective of the internal structure of the system, the value of the entropy in the equilibrium state tries to attain a maximum (bounded by the condition that the average energy is constant). In a sense, it is natural: random noise makes the system as disordered as possible. Let us now consider the form of the probability distribution function, which would maximize the entropy. To take into account the two constraints – the conservation of the average energy, Eq. (1.5), and the normalization $\sum_{\alpha} P_{\alpha} = 1$ – one can use the method of the Lagrange multipliers. Therefore, the following expression must be maximized with respect to all possible distributions P_{α}:

$$S_{\beta,\gamma}[P] = -\sum_{\alpha} P_{\alpha} \log(P_{\alpha}) - \beta \left(\sum_{\alpha} P_{\alpha} H_{\alpha} - E \right) - \gamma \left(\sum_{\alpha} P_{\alpha} - 1 \right) \tag{1.6}$$

where β and γ are the Lagrange multipliers. Variation with respect to P_{α} gives:

$$P_{\alpha} = \frac{1}{Z} \exp(-\beta H_{\alpha}) \tag{1.7}$$

where

$$Z = \sum_{\alpha} \exp(-\beta H_{\alpha}) = \exp(\gamma + 1) \tag{1.8}$$

is called the partition function, and the parameter β, which is called the inverse temperature, is defined by the condition:

$$\frac{1}{Z} \sum_{\alpha} H_{\alpha} \exp(-\beta H_{\alpha}) = E \tag{1.9}$$

In practice, however, it is the temperature that is usually taken as an independent parameter, whereas the average energy is obtained as the function of the temperature by Eq. (1.9).

The other fundamental quantity of the statistical mechanics is the free energy defined as follows:

$$F = E - TS \tag{1.10}$$

where $T = 1/\beta$ is the temperature. Using Eq.(1.7), one can easily derive the following basic relations among the free energy, the partition function, the entropy and the average energy:

$$F = -T\log(Z) \tag{1.11}$$

$$S = \beta^2 \frac{\partial F}{\partial \beta} \tag{1.12}$$

$$E = -\frac{\partial}{\partial \beta}\log(Z) = F + \beta\frac{\partial F}{\partial \beta} \tag{1.13}$$

Note, that according to the definition given by Eq. (1.10), the principle of the maximum of entropy is equivalent to that of the minimum of the free energy. One can easily confirm that taking the free energy (instead of the entropy) as the fundamental quantity which must be minimal with respect to all possible distribution functions, the same form of the probability distribution as given by Eq. (1.7) is obtained.

1.2 The mean-field approximation

In magnetic materials the microscopic state of the system is supposed to be defined by the values of the local spin magnetizations. In many magnetic materials the electrons responsible for the magnetic behavior are localized near the atoms of the crystal lattice, and the force that tends to orient the spins is the (short range) exchange interaction.

The most popular classical models, which describe this situation qualitatively, are called the Ising models. The microscopic variables in these systems are the Ising spins σ_i, which by definition can take only two values: +1 or −1. The traditional form for the microscopic energy (which from now on will be called the Hamiltonian) as the function of all the Ising spins is in the following:

$$H = -\sum_{<i,j>} J_{ij}\sigma_i\sigma_j - h\sum_i \sigma_i \tag{1.14}$$

Here the notation $<i,j>$ indicates the summation over all the lattice sites of the nearest neighbors, J_{ij} are the values of the spin–spin interactions, and h is the external magnetic field. If all the J_{ij}s are equal to a positive constant, then one gets the ferromagnetic Ising model, and if all the J_{ij}s are equal to a negative constant, then one gets the antiferromagnetic Ising model.

In spite of the apparent simplicity of the Ising model, an exact solution (which means the calculation of the partition function and the correlation functions) has been found only for the one- and the two-dimensional systems in the zero external magnetic field. In all other cases one needs to use approximate methods. One of the simplest methods is called the mean-field approximation. In many cases this method gives results that are not too far from the correct ones, and very often it is possible to get some qualitative understanding of what is going on in the system under consideration.

The starting point of the mean-field approximation is the assumption about the structure of the probability distribution function. It is assumed that the distribution function in the equilibrium state can be factorized as the product of the *independent* distribution functions in the lattice sites:

$$P[\sigma] = \frac{1}{Z}\exp(-\beta H[\sigma]) \simeq \prod_i P_i(\sigma_i) \tag{1.15}$$

The normalized site distribution functions take the form:

$$P_i(\sigma_i) = \frac{1+\phi_i}{2}\delta(\sigma_i - 1) + \frac{1-\phi_i}{2}\delta(\sigma_i + 1) \tag{1.16}$$

where ϕ_i are the parameters that have to be specified.

The factorization of the distribution function, Eq. (1.15), means that the average of any product of any functions at different sites is also factorizing on the product of the independent averages:

$$\langle f(\sigma_i)g(\sigma_j)\rangle = \langle f(\sigma_i)\rangle\langle g(\sigma_j)\rangle \tag{1.17}$$

where, according to the ansatz (1.15):

$$\langle f(\sigma_i)\rangle = \frac{1+\phi_i}{2}f(1) + \frac{1-\phi_i}{2}f(-1) \tag{1.18}$$

In particular, for the average site magnetizations, one easily gets:

$$\langle \sigma_i\rangle = \phi_i \tag{1.19}$$

Therefore, the physical meaning of the parameters $\{\phi_i\}$ in the trial distribution function is that they describe the average site spin magnetizations. According to the general principles of the statistical mechanics, these parameters must be such that they would minimize the free energy of the system.

Using Eqs. (1.15) and (1.16) for the entropy and for the average energy, one gets:

$$\begin{aligned} S &= -\langle \log(P[\sigma])\rangle \simeq -\sum_i \langle \log(P_i(\sigma_i))\rangle \\ &= -\sum_i \left[\frac{1+\phi_i}{2}\log\left(\frac{1+\phi_i}{2}\right) + \frac{1-\phi_i}{2}\log\left(\frac{1-\phi_i}{2}\right)\right] \end{aligned} \tag{1.20}$$

$$E = -\frac{1}{2}\sum_{<i,j>} J_{ij}\phi_i\phi_j - h\sum_i \phi_i \tag{1.21}$$

For the free energy, Eq. (1.10), one obtains:

$$F = -\frac{1}{2}\sum_{<i,j>} J_{ij}\phi_i\phi_j - h\sum_i \phi_i + T\sum_i \left[\frac{1+\phi_i}{2}\log\left(\frac{1+\phi_i}{2}\right) + \frac{1-\phi_i}{2}\log\left(\frac{1-\phi_i}{2}\right)\right] \tag{1.22}$$

To be more specific, consider the ferromagnetic system on the D-dimensional cubic lattice. In this case all the spin–spin couplings are equal to some positive constant: $J_{ij} = (1/2D)J > 0$, (the factor $1/2D$ is inserted just for convenience) and each site has $2D$ nearest neighbors. Since the system is homogeneous, it is natural to expect that all the ϕ_is must be equal to some constant ϕ. Then, for the free energy (1.22) one gets:

$$\frac{F}{V} \equiv f(\phi) = -\frac{1}{2}J\phi^2 - h\phi + T\left[\frac{1+\phi}{2}\log\left(\frac{1+\phi}{2}\right) + \frac{1-\phi}{2}\log\left(\frac{1-\phi}{2}\right)\right] \tag{1.23}$$

where V is the total number of sites (which is proportional to the volume of the system) and f is the density of the free energy. The necessary condition for the minimum of f is

$$\frac{df(\phi)}{d\phi} = 0$$

or:

$$-J\phi - h + T\,\mathrm{arctanh}(\phi) = 0 \tag{1.24}$$

The resulting equation, which defines the order parameter ϕ is:

$$\phi = \tanh[\beta(J\phi + h)] \tag{1.25}$$

Note, that the minimum of the free energy is conditioned by $d^2f/d\phi^2 > 0$. Using Eq. (1.24), this condition can be reduced to

$$\frac{1}{1-\phi^2} > \beta J \tag{1.26}$$

Consider first the case of a zero external magnetic field ($h = 0$). One can easily see that if $T > T_c = J$, the only solution of the Eq. (1.25) $\phi = \tanh(\beta J\phi)$ is $\phi = 0$, and this solution satisfies the condition (1.26). Therefore, at all temperatures higher than T_c the minimum of the free energy is achieved in the state in which all the site spin magnetizations are zero.

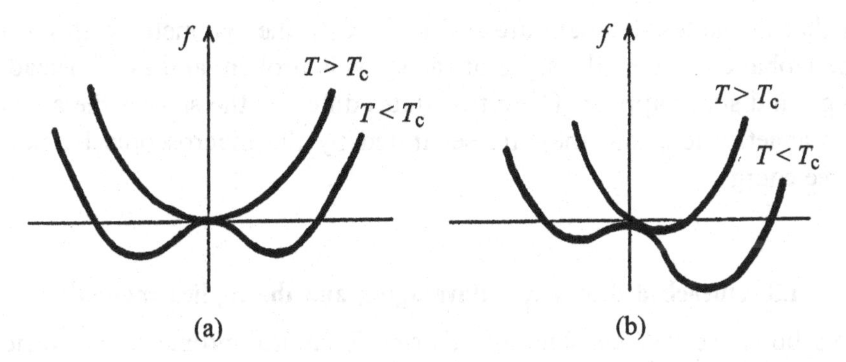

Fig. 1.1. Free energy of the ferromagnetic Ising magnet: (a) in the zero external magnetic field; (b) in non-zero magnetic field.

However, if $T < T_c$, then in addition to the solution $\phi = 0$, Eq. (1.25) (with $h = 0$) has two non-trivial solutions $\phi = \pm\phi(T) \neq 0$. One can easily check that in this temperature region the solution $\phi = 0$ becomes the maximum and not the minimum of the free energy, while the true minima are achieved at $\phi = \pm\phi(T)$. Therefore, in the low-temperature region $T < T_c$ the free energy has two minima, which are characterized by non-zero site magnetizations with opposite signs.

Near T_c the magnetization $\phi(T)$ is small. In this case the expansion in powers of ϕ in Eq. (1.25) can be made. In the leading order in $\tau \equiv (T/T_c - 1)$, $|\tau| \ll 1$, one gets:

$$\phi(T) = \text{const } |\tau|^{1/2}, \qquad (\tau < 0) \tag{1.27}$$

Thus, as $T \to T_c$ from below, $\phi(T) \to 0$. The expansion of the free energy Eq. (1.23) as the function of a small value of ϕ yields:

$$f(\phi) = \tfrac{1}{2}\tau\phi^2 + \tfrac{1}{4}g\phi^4 - h\phi \tag{1.28}$$

where $g = T/3$ and for simplicity we have taken $J = 1$. The qualitative shapes of $f(\phi)$ at $T > T_c$ $(\tau > 0)$ and at $T < T_c$ $(\tau < 0)$ are shown in Fig. 1.1. Note, that since the total free energy F is proportional to the volume of the system, the value of the free energy barrier separating the states with $\phi = \pm\phi(T)$ at $T < T_c$ is also proportional to the volume of the system. Therefore, in the *thermodynamic limit* $V \to \infty$ (which corresponds to the consideration of the *macroscopic* systems) the barrier separating the two states is becoming infinite.

The simple considerations described above demonstrate on a qualitative level the fundamental phenomenon called spontaneous symmetry breaking. At the temperature $T = T_c$, a phase transition of the second order occurs, such that in the low-temperature region $T < T_c$ the symmetry with respect to the global change of the signs of the spins is broken, and *two* (instead of one) ground states appear. These two states differ by the sign of the average spin magnetization, and they are separated by the macroscopic barrier of the free energy.

1.3 Quenched disorder, selfaveraging and the replica method

In this book we will consider the thermodynamical properties of various spin systems that are characterized by the presence of some kind of a quenched disorder. In realistic magnetic materials such disorder can exist, for example, owing to the oscillating nature of the exchange spin–spin interactions combined with the randomness in the positions of the interacting spins (such as in metallic spin-glass alloys AgMn), or owing to defects in the lattice structure, or because of the presence of impurities, etc.

Because we will mostly be interested in the qualitative effects produced by the quenched disorder, the details of the realistic structure of such magnetic systems will be left aside. Here we will concentrate on extremely simplified *model* description of the disordered spin systems.

In what follows we will consider two essentially different types of disordered magnet. First, we will study the thermodynamic properties of spin systems in which the disorder is strong. The term 'strong disorder' refers to the situation when the disorder appears to be the dominant factor for the ground-state properties of the system, so that it dramatically changes the low-temperature properties of the magnetic system as compared with the usual ferromagnetic phase. These types of system, usually called the *spin glasses*, will be considered in the first part of the course.

In the second part of the course we will consider the properties of weakly disordered magnets. This is the case when the disorder does not produce notable effects for the ground-state properties. It will be shown, however, that in certain cases even a small amount of disorder can produce dramatic effects for the critical properties of the system in close vicinity of the phase transition point.

The main problem in dealing with disordered systems is that the disorder here is *quenched*. Formally, all the results for the observable quantities for a given concrete system must depend on the concrete interaction matrix J_{ij}, i.e. these results are defined by a macroscopic number of random parameters. Apparently, results of this type are impossible to calculate and, moreover, they are useless. Intuitively it is clear, however, that the quantities which are called the observables should depend on some general averaged characteristics of the random interactions. This brings us to the concept of *selfaveraging*.

The traditional speculation about why the selfaveraging phenomenon should be expected to take place, is as follows. The free energy of the system is known to be proportional to the volume V of the system. Therefore, in the thermodynamic limit $V \to \infty$ the main contribution to the free energy comes from the volume, and not from the boundary, which usually produces effects of the next orders in the small parameter $1/V$. Any macroscopic system could be divided into a macroscopic number of macroscopic subsystems. Then the total free energy of the system would consist of the sum of the free energies of the subsystems, plus the contribution that comes from the interactions of the subsystems, at their boundaries. If all the interactions in the system are short range (which takes place in any realistic system), then the contributions from the mutual interactions of the subsystems are just the boundary effects, which vanish in the thermodynamic limit. Therefore, the total free energy could be represented as a sum of the macroscopic number of terms. Each of these terms is a random quenched quantity because it contains, as parameters, the elements of the random spin–spin interaction matrix. In accordance with the law of large numbers, the sum of many random quantities can be represented as their average value, obtained from their statistical distribution, multiplied by their number (all this is true, of course, only under certain requirements on the characteristics of the statistical distribution). Therefore, the total free energy of a macroscopic system must be selfaveraging over the *realizations* of the random interactions in accordance with their statistical distribution.

The free energy is known to be given by the logarithm of the partition function. Thus, in order to calculate the observable thermodynamics, one has to average the logarithm of the partition function over the given distribution of random J_{ij}s after the calculation of the partition function itself. To perform such a program the following technical trick, which is called the replica method, is used.

Formally, the replicas are introduced as follows. In order to obtain the physical (selfaveraging) free energy of the quenched random system we have to average the logarithm of the partition function:

$$F \equiv \overline{F_J} = -\frac{1}{\beta}\overline{\ln(Z_J)} \tag{1.29}$$

where $\overline{(\ldots)}$ denotes the averaging over random interactions $\{J_{ij}\}$ with a given distribution function $P[J]$:

$$\overline{(\dots)} \equiv \left(\prod_{<i,j>} \int dJ_{ij} \right) P[J](\dots) \tag{1.30}$$

and the partition function is

$$Z_J = \sum_{\sigma} \exp(-\beta H[J, \sigma]) \tag{1.31}$$

To perform this averaging procedure, the following trick is invented. Let us consider the *integer* power n of the partition function (1.31). This quantity is the partition function of the n non-interacting *identical* replicas of the original system (i.e. having identical fixed spin–spin couplings J_{ij}):

$$Z_J^n = \left(\prod_{a=1}^{n} \sum_{\sigma^a} \right) \exp\left(-\beta \sum_{a=1}^{n} H[J, \sigma_a] \right) \tag{1.32}$$

Here the subscript a labels the replicas. Let us introduce the quantity:

$$F_n = -\frac{1}{\beta n} \ln(Z_n) \tag{1.33}$$

where

$$Z_n \equiv \overline{Z_J^n} \tag{1.34}$$

Now, if a *formal* limit $n \to 0$ is taken in the expression (1.33), then the original expression for the physical free energy (1.29) will be recovered:

$$\lim_{n\to 0} F_n = -\lim_{n\to 0} \frac{1}{\beta n} \ln(Z_n) = -\lim_{n\to 0} \frac{1}{\beta n} \ln \left[\overline{\exp(n \ln Z_J)} \right] = -\frac{1}{\beta} \overline{\ln Z_J} = F \tag{1.35}$$

Thus, the scheme of the replica method can be described in the following steps. First, the quantity F_n for the integer n must be calculated. Second, the analytic continuation of the obtained function of the parameter n should be made for an arbitrary non-integer n. Finally, the limit $n \to 0$ has to be taken. Although this procedure may look rather doubtful at first, actually it is not so crazy! First, if the free energy appears to be an analytic function of the temperature and the other parameters (so that it can be represented as the series in powers of β), then the replica method can be easily proved to be correct in a strict sense. Second, in all cases, when the calculations can be performed by some other method, the results of the replica method are confirmed.

One could also introduce replicas another way [2, 29, 30]. Let us consider a general spin system described by a Hamiltonian $H[J;\sigma]$, which depends on the spin variables $\{\sigma_i\}$ and the spin–spin interactions J_{ij} (the concrete form of the Hamiltonian is irrelevant). If the interactions J_{ij} are quenched, the free energy of the system depends on the concrete realization of the J_{ij}s:

$$F[J] = -\frac{1}{\beta}\log(Z_J) \tag{1.36}$$

Now, let us assume that the spin–spin interactions are *partially annealed* (i.e. not perfectly quenched), so that they can also change their values, but the characteristic time scale of their changes is much larger than the time scale at which the spin degrees of freedom reach the thermal equilibrium. In this case the free energy given by (1.36) would still make sense, and it would become the energy function (the Hamiltonian) for the degrees of freedom of J_{ij}s.

Besides, the space in which the interactions J_{ij} take their values should be specified separately. The interactions J_{ij} could be discrete variables taking values $\pm J_0$, or they could be continuous variables taking values in some restricted interval, or they could be something else. In the quenched case this space of J_{ij} values is defined by a statistical distribution function $P[J]$. In the case of partial annealing, this function $P[J]$ has a meaning of the internal potential for the interactions J_{ij}, which restricts the space of their values.

Let us now assume that the spin and the interaction degrees of freedom *are not thermally equilibrated*, so that the degrees of freedom of the interactions have their own temperature T', which is different from the temperature T of the spin degrees of freedom. In this case for the total partition function of the system one gets:

$$\begin{aligned}\mathcal{Z} &= \int DJP[J]\exp(-\beta' F[J]) \\ &= \int DJP[J]\exp(\frac{\beta'}{\beta}\log Z_J) \\ &= \int DJP[J](Z_J)^n \equiv \overline{(Z_J)^n}\end{aligned} \tag{1.37}$$

where $n = T/T'$. Correspondingly, the total free energy of the system would be:

$$\mathcal{F} = -T'\log\left[\overline{(Z[J])^n}\right] \tag{1.38}$$

In this way we arrive at the replica formalism again, in which the 'number of replicas' $n = T/T'$ appears to be the *finite* parameter.

To obtain the physical (selfaveraging) free energy in the case of the quenched random J_{ij}s one takes the limit $n \to 0$. From the point of view of partial annealing, this situation corresponds to the limit of the infinite temperature T' in the system of J_{ij}s. This is natural in a sense that in this case the thermodynamics of the spin degrees of freedom produces no effect on the distribution of the spin–spin interactions.

In the case when the spin and the interaction degrees of freedom are thermally equilibrated, $T' = T$ $(n = 1)$, we arrive at the trivial case of the purely annealed disorder, irrespective of the difference between the characteristic time scales of the J_{ij} interactions and the spins. This is also natural because the thermodynamic description formally corresponds to the infinite times, and the characteristic time scales of the dynamics of the internal degrees of freedom become irrelevant. If $n \neq 0$ and $n \neq 1$, one gets the situation that could be called partial annealing, and which is the intermediate case between quenched disorder and annealed disorder.

Part one

Spin-glass systems

2

Physics of the spin-glass state

Before starting doing detailed calculations, first it would be useful to get a qualitative understanding of the general physical phenomena taking place in statistical mechanics of spin systems with strong quenched disorder. Therefore, in this chapter we will discuss the problem of spin-glass state only in simple qualitative terms.

2.1 Frustrations

There are quite a few statistical models of spin glasses. Here we will concentrate on one of the simplest models, which can be formulated in terms of the classical Ising spins, described by the following Hamiltonian:

$$H = -\frac{1}{2}\sum_{i \neq j}^{N} J_{ij}\sigma_i\sigma_j \tag{2.1}$$

This system consists of N Ising spins $\{\sigma_i\}$ $(i = 1, 2, \ldots, N)$, taking values ± 1 which are placed in the vertices of some lattice. The spin–spin interactions J_{ij} are random in their values and signs. The properties of such a system are defined by the statistical distribution function $P[J_{ij}]$ of the spin–spin interactions. For the moment, however, the concrete form of this distribution will not be important. The motivation for the Hamiltonian (2.1) from the point of view of realistic spin-glass systems is well described [4].

The crucial phenomena revealed by a strong quenched disorder, which make such systems so hard to study, are as follows. Consider the system of three interacting spins (Fig. 2.1). Let us assume for simplicity that the interactions among them can be different only in their signs, being equal in the absolute value. Then for the ground state of such a system we can find two essentially different situations.

If all three interactions J_{12}, J_{23} and J_{13} are positive, or two of them are

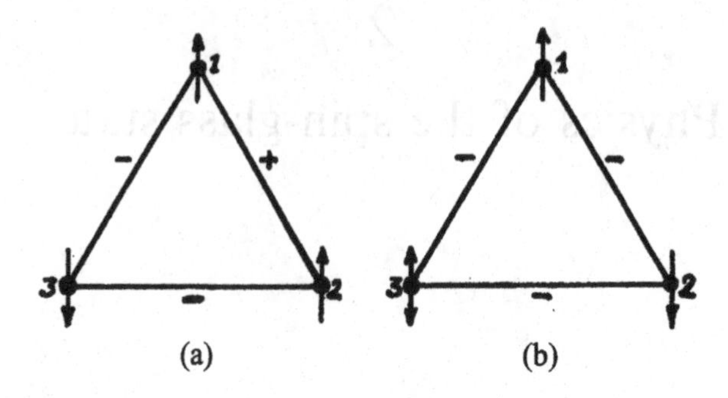

Fig. 2.1. The frustrations in the system of three spins. (a) No frustration: the product of the interactions along the triangle is positive. (b) The frustated triangle: the product of the interactions along the triangle is negative.

negative while the third one is positive, then the ground state of this three-spin system is unique (except for the global change of signs of all the spins, Fig. 2.1(a)). This is the case when the product of the interactions along the triangle is positive.

However, if the product of the interactions along the triangle is negative (one of the interactions is negative, or all three interactions are negative, Fig. 2.1(b)), then the ground state of such a system is degenerate. One can easily check, going from spin to spin along the triangle, that in this case the orientation ('plus' or 'minus') of one of the spins remains 'unsatisfied' with respect to the interactions with its neighbors.

One can also easily check that a similar phenomenon takes place in any closed spin chain of arbitrary length, provided that the product of the spin–spin interactions along the chain is negative. This phenomenon is called *frustration*† [17].

One can easily see that not *any* disorder induces frustrations. On the other hand, it is the frustrations that describe the relevant part of the disorder, and which essentially affect the ground-state properties of the system. In other words, if the disorder does not produce frustrations, it can be considered as being irrelevant. In some cases an irrelevant disorder can just be removed by a proper redefinition of the spin variables of the system. A simple example of this situation is illustrated by the so-called Mattice magnet. This also is a formally disordered spin system, which is described by the Hamiltonian (2.1), where the spin–spin interactions are defined as $J_{ij} = \xi_i \xi_j$, and the quenched ξ_is take values ± 1 with equal probability. In such a system the

† This term is quite adequate in its literal meaning, as the triangle discussed above might as well be interpreted as the famous love triangle. Besides, the existence of frustrations in spin glasses removes any hope for finding a simple solution of the problem.

interactions J_{ij} are also random in signs, although one can easily check that no frustrations appear here. Moreover, after simple redefinition of the spin variables, $\sigma_i \to \sigma_i \xi_i$, an ordinary ferromagnetic Ising model will be recovered. Thus, this type of disorder (called the Mattice disorder) is actually fictitious for the thermodynamic properties of the system.

It is crucial that the 'true' disorder with frustrations can not be removed by any transformation of the spin variables. Because in a macroscopic spin system, in general, one can draw a lot of different frustrated closed spin chains, the total number of frustrations must also be macroscopically large. This, in turn, would result either in a tremendous degeneracy of the ground state, or, in general, it could produce a lot of low-lying states with the energies very close to the ground state.

2.2 Ergodicity breaking

Formally, according to the general selfaveraging arguments (Section 1.3), to derive the observable thermodynamics of a disordered spin system one has to find a way for averaging the logarithm of the partition function over random parameters J_{ij} simultaneously with the calculation of the partition function itself. It is clear that this problem is not easy, but nevertheless it looks just like a technical problem (well, presumably a very hard one), and not more than that. Actually, for spin-glass type systems this is not just a *technical* problem. To realize this, let us consider again a few general points of statistical mechanics.

Everything would be rather simple if the free energy in the thermodynamic limit was an analytic function of the temperature and the other parameters. Actually, for most of non-trivial systems that are of interest in statistical mechanics, this is not so. Very often, owing to spontaneous breaking of some kind of a symmetry in the thermodynamic limit, there exists a *phase transition*, and this makes the free energy a *non-analytic* function of the parameters involved.

Let us consider again the ordinary ferromagnetic Ising model (Chapter 1), which in very simple terms illustrates the physical consequences of this phenomenon. Because the Hamiltonian of this system is invariant with respect to the global change of the signs of all the spins, any thermodynamic quantity that is odd in spins must be identically equal to zero. In particular, this must be true for the quantity that describes the global magnetization of the system. If the volume N of the system is *finite*, these arguments are indeed perfectly correct. However, in the thermodynamic limit $N \to \infty$ we are facing a rather non-trivial situation. According to simple calculations performed in Section 1.2, the free energy as a function of the global magnetization acquires the double-well shape (Fig. 1.1) at low temperatures. The value of the energy barrier separating the two ground states is proportional to the volume of the system, and it is getting infinite in the limit $N \to \infty$. In other words, at temperatures below T_c the space of all microscopic states of the system is divided into two equal valleys separated by the infinite barrier. On the other hand, according to the fundamental ergodic hypothesis of statistical mechanics (Section 1.1), it is assumed that in the limit of infinite observation time the system (following its internal dynamics) visits all its microscopic states many times, and it is this assumption that makes it possible to apply the statistical mechanical approach: for the calculation of the averaged quantities we use averaging over the ensemble of states with the corresponding probability distribution instead of that over time. In the situation under consideration, when the thermodynamic limit $N \to \infty$ is taken *before* the observation time goes to infinity (it is this order of limits that corresponds to the adequate statistical mechanical description of a *macroscopic* system), the above ergodic assumption simply does not work. Whatever the (reasonable) internal dynamics of the system, it could never make it possible to jump over the infinite energy barrier separating the two valleys of the space of states. Thus, in the *observable* thermodynamics, only half of the states contribute (these are the states that are on one side from the barrier), and that is why in the observable thermodynamics the global magnetization of the system appears to be non-zero.

In the terminology of statistical mechanics, this phenomenon is called *ergodicity breaking*, and it manifests itself as *spontaneous symmetry breaking*: below T_c the observable thermodynamics becomes non-symmetric with respect to the global change of signs of all the spins. As a consequence, in the calculations of the partition function below T_c, one has to take into account not all, but only one half of all the microscopic states of the system (the states that belong to one valley).

The above example of the ferromagnetic system is very simple because here one can easily guess right away what kind of symmetry could be broken at low temperatures. In spin glasses, spontaneous symmetry breaking also takes place. However, unlike the ferromagnetic system, here it is much more difficult to guess which one. The main problem is that the symmetry which might be broken in a given sample can depend on the quenched disorder parameters involved. In this situation the calculation of the *observable* free energy becomes an extremely difficult problem, because now one must take into account only the states belonging to one of the many valleys, while the structure of these valleys depends on a concrete realization of the random disorder parameters.

2.3 Continuous sequence of phase transitions

Of course, the existence of many local minimum states in the frustrated spin system does not automatically means that at low temperatures these states create their valleys separated by the infinite barriers of the free energy. Because of thermal fluctuations (which are usually rather strong in the low-dimensional systems) the energy barriers could effectively 'melt', and in this case the ground state of the free energy could appear to be unique. Then there will be no spontaneous symmetry breaking, and at any finite temperature the system will be in the 'symmetric' paramagnetic state. Of course, from the point of view of the anomalously slow dynamic relaxation properties this state can be essentially different from the usual high-temperature paramagnetic state, but this problem would lead us well beyond the scope of pure statistical mechanics.

It could also happen that because of some symmetry properties the global minimum of the free energy of a given sample is achieved at low temperatures at some unique non-trivial spin configuration (of course, in this case the 'counterpart' spin state that differs by the global change of the signs of the spins must also be the ground state). It would mean that at low enough temperatures (below a certain phase-transition temperature T_c), the system must 'freeze' in this unique random spin state, which will be characterized by the non-zero values of the thermally averaged local spin magnetizations at each site $\langle\sigma_i\rangle$. As this ground state is random, the values of the local magnetizations $\langle\sigma_i\rangle$ will fluctuate in their values and signs from site to site, so that the usual ferromagnetic order parameter, which describes the global magnetization of the system, $m = 1/N \sum_i \langle\sigma_i\rangle$, must be zero (in the infinite volume limit). However, this state can be characterized by the other thermodynamic order parameter (usually called the Edwards–Anderson order parameter [18]):

$$q = \frac{1}{N} \sum_i \langle\sigma_i\rangle^2 \neq 0 \tag{2.2}$$

The properties of systems of this type are studied in detail in the papers by Fisher and Huse [7], and we will not consider them here.

In the subsequent chapters we will concentrate on a qualitatively different situation, which arises when there exist *macroscopically* large numbers of states in which the system could get 'frozen' at low temperatures. Moreover, unlike 'ordinary' statistical mechanical systems, according to the mean-field theory of spin glasses the spontaneous symmetry breaking in the spin-glass state takes place not just at a certain T_c, but it occurs at *any* temperature below T_c. In other words, below T_c a *continuous sequence* of the phase

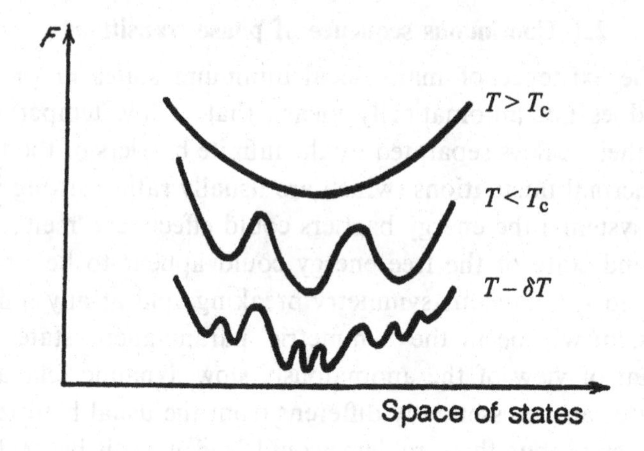

Fig. 2.2. The qualitative structure of the spin glass free-energy landscape at different temperatures.

transitions takes place and, correspondingly, the free energy appears to be non-analytic at any temperature below T_c.

In general qualitative terms this phenomenon can be described as follows. Just below a certain critical temperature T_c, the space of spin states is divided into *many* valleys (their number diverges in the thermodynamic limit), separated by infinite barriers of the free energy. At the temperature $T = T_c - \delta T$ each valley is characterized by the non-zero values of the average local spin magnetizations $\langle \sigma_i \rangle_{(\alpha)}$ (which, of course, fluctuate in sign and magnitude from site to site). Here $\langle \ldots \rangle_\alpha$ denotes the thermal average inside a particular valley α. The order parameter, which describes the degree of freezing of the system inside the valleys, could be defined as follows:

$$q(T) = \frac{1}{N} \sum_i \langle \sigma_i \rangle^2_{(\alpha)} \tag{2.3}$$

According to the mean-field theory of spin glasses, the value of q depends only on the temperature, and it appears to be the same for all the valleys. At $T \to T_c$, $q(T) \to 0$.

At a further decrease of the temperature, new phase transitions of ergodicity breaking take place, so that each valley splits into many new ones separated by infinite barriers of free energy (Fig. 2.2). The state of the system in all new valleys can again be characterized by the order parameter (2.3), and its value grows as the temperature decreases.

As the temperature goes down to zero, this process of fragmentation of the space of states into smaller and smaller valleys goes on *continuously*. In a sense, it means that at *any* temperature below T_c the system is in the critical state.

To what extent this situation is realistic from the experimental point of view remains open, although the series of recent experiments (which will be discussed in Chapter 9) gives a strong indication in favor of it. In any case, this new type of physics is very interesting in itself, and it is worthy of study.

2.4 Order parameter

It is clear that the order parameter (2.3) defined for one valley only does not contain any information about the other valleys, and it does not tell us anything about the structure of space of the ground states as a whole. Let us try to construct the other physical order parameter, which would describe this structure as fully as possible.

Consider the following series of imaginary experiments. Let us fix an arbitrary spin state, and then at a given temperature T below T_c let the system relax to thermal equilibrium. For each experiment a new starting random spin state should be taken. Then each experiment will be characterized by some equilibrium values of the average local spin magnetizations $\langle\sigma_i\rangle_{(\alpha)}$, where α denotes the number of the experiment. Since there exists a macroscopically large number of valleys in the phase space in which the system could get 'trapped', these site magnetizations, in general, could be different for different experiments.

Let us assume that we have performed an infinite number of such experiments. Then, we can introduce the quantity that would describe to what extent the states which have been obtained in different experiments are close to each other:

$$q_{\alpha\beta} = \frac{1}{N}\sum_{i}^{N}\langle\sigma_i\rangle_{(\alpha)}\langle\sigma_i\rangle_{(\beta)} \tag{2.4}$$

It is clear that $|q_{\alpha\beta}| \leq 1$, and the maximum value of $q_{\alpha\beta}$ is achieved when the two states in the experiments α and β coincide (in this case, the overlap (2.4) coincides with that of (2.3), which has been introduced for one valley only). It is also clear that the less correlated the two different states are, the smaller the value of the overlap (2.4) they have. If the two states are not correlated at all, then their overlap (in the thermodynamic limit) is equal to zero. In a sense the overlap $q_{\alpha\beta}$ defines a kind of a metric in the space of states (the quantity $q_{\alpha\beta}^{-1}$ could be conditionally called the 'distance' in the space of states).

To describe the statistical properties of these overlaps one can introduce the following probability distribution function:

$$P(q) = \sum_{\alpha\beta} \delta(q_{\alpha\beta} - q) \quad (2.5)$$

It appears that it is in terms of this distribution function $P(q)$ that the spin-glass state looks essentially different from any other 'ordinary' thermodynamic state.

Possible types of the function $P(q)$ are shown in Fig. 2.3. The paramagnetic phase is characterized by the only global minimum of the free energy, in which all the site magnetizations are equal to zero. Therefore, the distribution function $P(q)$ in this phase is the δ-function at $q = 0$ (Fig. 2.3(a)). In the ferromagnetic phase there exist two minima of the free energy with the site magnetizations $\pm m$. Thus, the distribution function $P(q)$ in this phase must contain two δ-peaks at $q = \pm m^2$ (Fig 2.3(b)). It is clear that in the case of the 'fake' spin-glass phase in which there exist only two disordered global minima (the states that differ by the global reversal of the local spin magnetizations) the distribution function $P(q)$ must look the same as in the ferromagnetic state.

According to the mean-field theory of spin glasses, which will be considered in the subsequent chapters, the distribution function $P(q)$ in the 'true' spin-glass phase looks essentially different (Fig. 2.3(c)). Here, between the two δ-peaks at $q = \pm q_{\max}(T)$ there is a continuous curve. The value of $q_{\max}(T)$ is equal to the maximum possible overlap of the two ground states, which is the 'selfoverlap' (2.3). Because the number of valleys in the system is macroscopically large and their selfoverlaps are all equal, the function $P(q)$ has two δ-peaks at $q = \pm q_{\max}(T)$. The existence of the continuous curve in the interval $(0, \pm q_{\max}(T))$ is the direct consequence of the 'origin' of the spin states involved: because they appear as the result of a continuous process of fragmentation of the valleys into the smaller and smaller ones, the states that form such a type of *hierarchy* are necessarily correlated.

Thus, it is the distribution function $P(q)$ that can be considered as the proper physical order parameter, adequately describing the peculiarities of the spin-glass phase. Although the procedure of its definition described above looks somewhat artificial, it will be shown later that the distribution function $P(q)$ can be defined as the thermodynamical quantity and, moreover, in terms of the mean-field theory of spin glasses it can be calculated explicitly.

2.5 Ultrametricity

According to the qualitative picture described above, the spin-glass states are organized in a kind of a hierarchical structure (Fig. 2.2). It can be proved

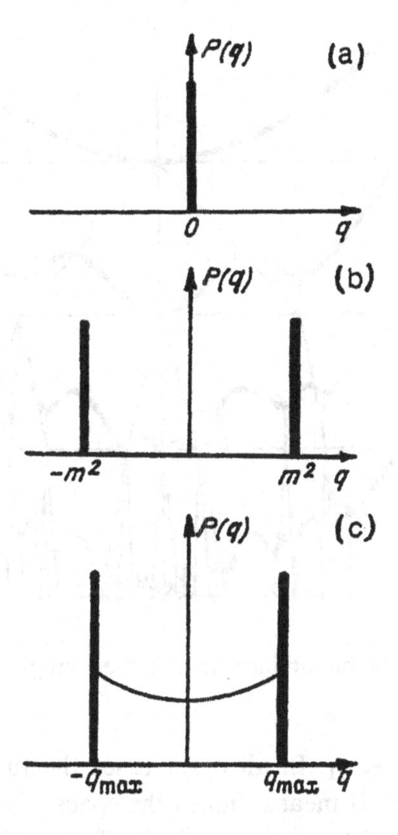

Fig. 2.3. The probability distribution function $P(q)$: (a) in the paramagnetic phase; (b) in the ferromagnetic phase; (c) in the spin-glass phase.

that this rather sophisticated space of states could be described in terms of well-defined thermodynamical quantities.

In the previous section we introduced the distribution function $P(q)$, which gives the probability of finding two spin-glass states having an overlap equal to q. Now let us introduce a somewhat more complicated distribution function $P(q_1, q_2, q_3)$, which gives the probability for *three* arbitrary spin-glass states to have their overlaps equal to q_1, q_2 and q_3:

$$P(q_1, q_2, q_3) = \sum_{\alpha\beta\gamma} \delta(q_{\alpha\beta} - q_1)\delta(q_{\alpha\gamma} - q_2)\delta(q_{\beta\gamma} - q_3) \tag{2.6}$$

In terms of the mean-field theory this function can also be calculated explicitly (Chapter 5). It can be shown that the function $P(q_1, q_2, q_3)$ is not equal to zero only if at least two of the three overlaps are equal to each other and their value is not larger than the third one. In other words, the function $P(q_1, q_2, q_3)$ is non-zero only in the following three cases: $q_1 = q_2 \leq q_3$;

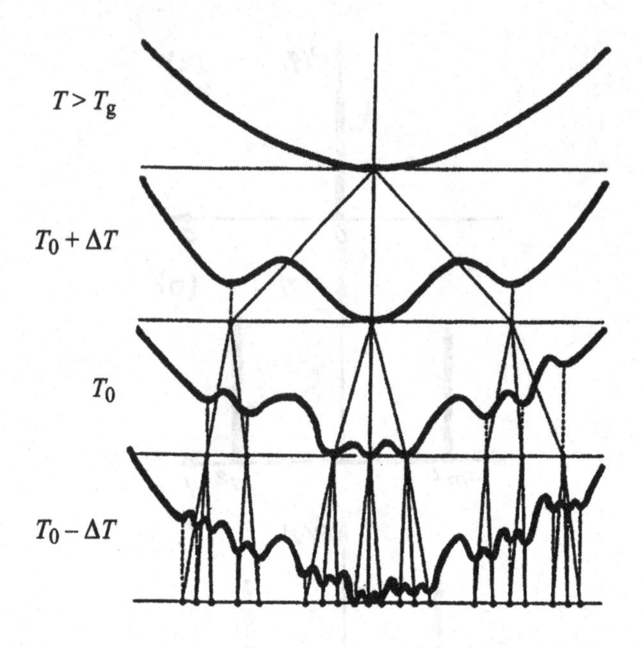

Fig. 2.4. The hierarchical tree of the spin-glass states.

$q_1 = q_3 \leq q_2$; $q_3 = q_2 \leq q_1$. In all other cases the function $P(q_1, q_2, q_3)$ is identically equal to zero. It means that in the space of spin-glass states there exist no triangles with all three sides being different. The spaces having the above metric property are called *ultrametric*. Ultrametricity from the point of view of physics (in mathematics, ultrametric structures have been known since the end of the nineteenth century) is described in detail in the review by Rammal *et al.* [3].

The simplest illustration of the ultrametric structure can be made in terms of the hierarchical tree (Fig. 2.4). Here the space of the spin-glass states is identified with the set of the endpoints of the tree. The metric in this space is defined in such a way that the overlap (the distance) between any two states depends only on the number of generations to their closest 'ancestor' on the tree (as the number of the generations increases, the value of the overlap decreases). One can easily check that the space with such metrics is ultrametric.

In the mean-field theory of spin glasses, such an illustrative tree of states actually describes the hierarchical fragmentation of the space of the spin-glass states into the valleys, as has been described above (Fig. 2.2). If for the vertical axis in Fig. 2.4 we assign the (discrete) value of the overlaps q, then the set of spin-glass states at any given temperature $T < T_c$ can be obtained at the cross-section of the tree at the level $q = q_{max}(T)$. After decreasing the temperature to a new value $T' < T$, each of the states at the level $q_{max}(T)$ gives birth to numerous 'descendants', which are the endpoints of the tree at the new level $q_{max}(T') > q_{max}(T)$. Correspondingly, after increasing the temperature to a higher value $T'' > T$, all the states having their common ancestors at the level $q_{max}(T'') < q_{max}(T)$ merge together into one state. As $T \to T_c$, $q_{max}(T) \to 0$, which is the level of the (paramagnetic) 'grandancestor' of all the spin-glass states.

Because the function $q_{max}(T)$ is determined by the temperature, it means that it is the temperature that defines the level of the tree at which the 'horizontal' cross-section should be made, and this, in turn, reveals all the spin-glass states at this temperature. All the states that are below this level are 'indistinguishable', while all the states that are above this level form the 'evolution history' of the spin-glass states at a given temperature. In this sense the temperature defines the elementary ('ultraviolet') *scale* in the space of the spin-glass states. This creates a kind of scaling in the spin-glass phase: by changing the temperature one just changes the scale in the space of the spin-glass states.

3

Mean-field theory of spin glasses

3.1 Infinite range interaction model

The Sherrington and Kirkpatric (SK) model of spin glasses [19] is defined by the usual Ising spins Hamiltonian:

$$H = -\sum_{i<j}^{N} J_{ij}\sigma_i\sigma_j \tag{3.1}$$

where the spin–spin interactions J_{ij} are random quenched variables that are described by the symmetric Gaussian distribution *independently* for any pair of sites (i, j):

$$P[J_{ij}] = \prod_{i<j}\left[\sqrt{\frac{N}{2\pi}}\exp\left(-\frac{N}{2}J_{ij}^2\right)\right] \tag{3.2}$$

According to the above definition, each spins interacts with all the other spins of the system. For that reason the space structure (dimensionality, type of the lattice, etc.) of this model is irrelevant for its properties. The space here is just the set of N sites in which the Ising spins are placed, and all these spins, in a sense, could be considered as the nearest neighbors. In the thermodynamic limit ($N \to \infty$) such a structure can be interpreted as the infinite dimensional lattice, and it is this property that makes the mean-field approach be exact.

According to the probability distribution (3.2) one gets:

$$\overline{J_{ij}} = 0; \qquad \overline{J_{ij}^2} = \frac{1}{N} \tag{3.3}$$

where $\overline{(\ldots)}$ denotes the averaging over random J_{ij}s:

$$\overline{(\ldots)} \equiv \int \mathbf{DJ}P[\mathbf{J}](\ldots) = \prod_{i<j}\left[\sqrt{\frac{N}{2\pi}}\int_{-\infty}^{+\infty} dJ_{ij}\exp\left(-\frac{N}{2}J_{ij}^2\right)\right](\ldots) \tag{3.4}$$

One can easily check that because of the chosen normalization of the order of $1/N$ for the average square values of the couplings J_{ij}, the average energy of the system appears to be of the order of N, as it should be for an adequately defined physical system.

It is clear, of course, that the microscopic structure of the model defined above is completely unphysical. Nevertheless, this model has two big advantages: first, it is exactly solvable; and second, its solution appears to be quite non-trivial. Moreover, on a qualitative level the physical interpretation of this solution can also be generalized for 'normal' random physical systems. If it were discovered (e.g. in experiments) that real spin glasses demonstrate the physical properties predicted from the solution of the SK model, then, in a sense, the original artificial system that initiated the true result would not be so important.

3.2 Replica symmetric solution

To calculate the replica free energy F_n, (Eq. (1.33)), according to Eqs. (1.32)–(1.34) one has to calculate the annealed average of the n-th power of the partition function:

$$Z_n = \sum_{\sigma_i^a} \int DJ_{ij} \exp\left(\beta \sum_{a=1}^{n} \sum_{i<j}^{N} J_{ij}\sigma_i^a\sigma_j^a - \frac{1}{2}N \sum_{i<j} J_{ij}^2\right) \tag{3.5}$$

(here and in what follows irrelevant pre-exponential factors are omitted). Integration over the J_{ij}s gives:

$$Z_n = \sum_{\sigma_i^a} \exp\left[\frac{\beta^2}{2N} \sum_{i<j}^{N} \left(\sum_{a=1}^{n} \sigma_i^a\sigma_j^a\right)^2\right] \tag{3.6}$$

or:

$$Z_n = \sum_{\sigma_i^a} \exp\left[\frac{1}{4}\beta^2 Nn + \frac{1}{2}\beta^2 N \sum_{a<b}^{n} \left(\frac{1}{N} \sum_{i}^{N} \sigma_i^a\sigma_i^b\right)^2\right] \tag{3.7}$$

The summation over the sites in the above equation can be linearized by introducing the replica matrix Q_{ab}:

$$Z_n = \left(\prod_{a<b}^{n} \int dQ_{ab}\right) \sum_{\sigma_i^a} \exp\left[\frac{1}{4}\beta^2 Nn - \frac{1}{2}\beta^2 N \sum_{a<b}^{n} Q_{ab}^2 + \beta^2 \sum_{a<b}^{n} \sum_{i}^{N} Q_{ab}\sigma_i^a\sigma_i^b\right] \tag{3.8}$$

The replica variables Q_{ab} have a clear physical interpretation. According to the above equation, the equilibrium values of the matrix elements Q_{ab} are defined by the equations $\delta Z_n/\delta Q_{ab} = 0$, which give:

$$Q_{ab} = \frac{1}{N}\sum_i^N \langle \sigma_i^a \sigma_i^b \rangle \tag{3.9}$$

Because the expression in the exponential of Eq. (3.8) is linear in the spatial summation, the partition function Z_n can be factorized into the independent site partition functions:

$$Z_n = \left(\prod_{a<b}^n \int dQ_{ab}\right) \exp\left[\frac{1}{4}\beta^2 Nn - \frac{1}{2}\beta^2 N \sum_{a<b}^n Q_{ab}^2\right] \times \prod_i^N \left[\sum_{\sigma_i^a} \exp\left(\beta^2 \sum_{a<b}^n Q_{ab}\sigma_i^a\sigma_i^b\right)\right] \tag{3.10}$$

or

$$Z_n = \left(\prod_{a<b}^n \int dQ_{ab}\right) \exp\left[\frac{1}{4}\beta^2 Nn - \frac{1}{2}\beta^2 N \sum_{a<b}^n Q_{ab}^2 + N\log\left[\sum_{\sigma_a} \exp\left(\beta^2 \sum_{a<b}^n Q_{ab}\sigma_a\sigma_b\right)\right]\right] \tag{3.11}$$

This equation can be represented as follows:

$$Z_n = \int D\hat{Q} \exp\left(-\beta n N f_n[\hat{Q}]\right) \tag{3.12}$$

where

$$f_n[\hat{Q}] = -\frac{1}{4}\beta + \frac{1}{2n}\beta \sum_{a<b}^n Q_{ab}^2 - \frac{1}{\beta n}\log\left[\sum_{\sigma_a} \exp\left(\beta^2 \sum_{a<b}^n Q_{ab}\sigma_a\sigma_b\right)\right] \tag{3.13}$$

In the thermodynamic limit the integral for the partition function (3.12) in the leading order in N is given by the saddle point of the function $f[\hat{Q}]$:

$$Z_n \simeq \left[\det \frac{\delta^2 f}{\delta \hat{Q}^2}\right]^{(-1/2)} \exp\left(-\beta n N f[\hat{Q}^*]\right) \tag{3.14}$$

Here $\hat{Q}^*$ is the matrix corresponding to the minimum of the function f, and it is defined by the saddle-point equation:

$$\frac{\delta f}{\delta Q_{ab}} = 0 \tag{3.15}$$

According to the general scheme of the replica method, the quantity $f[\hat{Q}^*]$ in the limit $n \to 0$ gives the density of the free energy of the system.

Thus, our strategy should continue as follows. First, for an arbitrary matrix $\hat{Q}$ one has to calculate an explicit expression for the replica free energy (3.13). Then one has to find the solution $\hat{Q}^*$ of the saddle-point equations (3.15) and the corresponding value for the replica free energy $f_n[\hat{Q}^*]$. Finally the limit $\lim_{n\to 0} f_n[\hat{Q}^*]$ should be taken. Unfortunately, this systematic program can not be fulfilled, because for an *arbitrary* matrix $\hat{Q}$ the replica free energy (3.13) cannot be calculated.

Therefore, the procedure of solving the problem must be somewhat more intuitive. First, one has to *guess* the correct structure of the solution $\hat{Q}^*$, which would hopefully depend on a limited number of parameters, and which would make possible the calculation of the replica free energy (3.13). Then these parameters should be obtained from the saddle-point equations (3.15), and finally the corresponding value of the saddle-point free energy should be calculated. Of course, according to this scheme, one would be able to find the extremum only inside some limited subspace of all matrices $\hat{Q}$. However, if it were possible to prove that the corresponding Hessian $\delta^2 f/\delta \hat{Q}^2$ at this extremum has been positively defined, then it would mean that the true extremum has been found. (Of course, this scheme does not guarantee that no other saddle points exist.)

Because all the replicas in our system are equivalent, one could naively guess that the adequate form of the matrix $\hat{Q}^*$ is such that all its elements are equal:

$$Q_{ab} = q; \qquad \text{for all } a \neq b \tag{3.16}$$

This ansatz is called the replica symmetric (RS) approximation. All the calculations in the RS approximation are very simple. For the replica free energy (3.13) one gets:

$$f(q) = -\frac{1}{4}\beta + \frac{\beta}{4n}n(n-1)q^2 - \frac{1}{\beta n}\log\left[\sum_{\sigma_a}\exp\left(\frac{1}{2}\beta^2(\sum_a^n \sigma_a)^2 q - \frac{1}{2}\beta^2 nq\right)\right] \tag{3.17}$$

In the standard way of introducing the Gaussian integration, one makes the quadratic term in the exponential to be linear in σ:

$$f(q) = -\frac{1}{4}\beta + \frac{1}{2}\beta q + \frac{1}{4}(n-1)\beta q^2$$
$$-\frac{1}{\beta n}\log\left[\int_{-\infty}^{+\infty}\frac{dz}{\sqrt{2\pi}}\exp\left(-\frac{1}{2}z^2\right)\sum_{\sigma_a=\pm 1}\exp\left(\beta z\sqrt{q}\sum_a^n \sigma_a\right)\right] \tag{3.18}$$

Summing over the σs one gets:

$$f(q) = -\frac{1}{4}\beta + \frac{1}{2}\beta q + \frac{1}{4}(n-1)\beta q^2$$
$$-\frac{1}{\beta n}\log\left(\int_{-\infty}^{+\infty}\frac{dz}{\sqrt{2\pi}}\exp\left(-\frac{1}{2}z^2\right)[2\cosh(\beta z\sqrt{q})]^n\right) \qquad (3.19)$$

The limit $n \to 0$ in the last term of the above equation is taken in the following way:

$$\lim_{n\to 0}\frac{1}{\beta n}\log\left(\int_{-\infty}^{+\infty}\frac{dz}{\sqrt{2\pi}}\exp\left(-\frac{1}{2}z^2\right)[2\cosh(\beta z\sqrt{q})]^n\right)$$
$$= \lim_{n\to 0}\frac{1}{\beta n}\log\left(\int_{-\infty}^{+\infty}\frac{dz}{\sqrt{2\pi}}\exp\left(-\frac{1}{2}z^2\right)\exp\{n\ln[2\cosh(\beta z\sqrt{q})]\}\right)$$
$$= \lim_{n\to 0}\frac{1}{\beta n}\log\left(\int_{-\infty}^{+\infty}\frac{dz}{\sqrt{2\pi}}\exp\left(-\frac{1}{2}z^2\right)\{1+n\ln[2\cosh(\beta z\sqrt{q})]\}\right)$$
$$= \lim_{n\to 0}\frac{1}{\beta n}\log\left(1+n\int_{-\infty}^{+\infty}\frac{dz}{\sqrt{2\pi}}\exp\left(-\frac{1}{2}z^2\right)\ln[2\cosh(\beta z\sqrt{q})]\right)$$
$$= \frac{1}{\beta}\int_{-\infty}^{+\infty}\frac{dz}{\sqrt{2\pi}}\exp\left(-\frac{1}{2}z^2\right)\ln[2\cosh(\beta z\sqrt{q})] \qquad (3.20)$$

Thus, for the density of the free energy (3.19) in the limit $n \to 0$ we finally obtain:

$$f(q) = -\frac{1}{4}\beta(1-q)^2 - \frac{1}{\beta}\int_{-\infty}^{+\infty}\frac{dz}{\sqrt{2\pi}}\exp\left(-\frac{1}{2}z^2\right)\ln[2\cosh(\beta z\sqrt{q})] \qquad (3.21)$$

Now from the condition $\partial f/\partial q = 0$ one can easily derive the corresponding saddle-point equation for the parameter q:

$$q = \int_{-\infty}^{+\infty}\frac{dz}{\sqrt{2\pi}}\exp\left(-\frac{1}{2}z^2\right)\tanh^2(\beta z\sqrt{q}) \qquad (3.22)$$

One can easily check that at $T \geq T_c = 1$ the only solution of this equation is $q = 0$. On the other hand, at $T < T_c$ there exists the non-trivial solution $q(T) \neq 0$. In the vicinity of the critical temperature, at $(1-T) \equiv \tau \ll 1$, this solution can be found explicitly: $q(\tau) = \tau + O(\tau^2)$. It is also easy to check that in the low-temperature limit $T \to 0$, $q(T) \to 1$.

According to Eqs. (3.16) and (3.9), the obtained solution for $q(T)$ gives us the physical order parameter:

$$q(T) = \frac{1}{N}\sum_i^N \langle\sigma_i\rangle^2 \qquad (3.23)$$

As $q(T)$ is not equal to zero in the low-temperature region, $T < T_c$, the spins of the system are getting frozen in some random state. Besides, as only one solution exists for $q(T)$, such a disordered ground state must be unique.

Using the result obtained for the free energy, one can easily perform further straightforward calculations to find all the other observable thermodynamical quantities, such as specific heat, susceptibility, entropy etc. Thus, in terms of the considered replica symmetric ansatz, a complete solution of the problem can easily be obtained.

All that would be very nice, if it were correct. Unfortunately, it is not. One of the simplest ways to see that there is something fundamentally wrong in the obtained solution is to calculate the entropy. One can easily check that at sufficiently low temperatures the entropy is becoming negative! (At $T = 0$ the entropy $S = -1/2\pi \simeq -0.17$.) Moreover, the calculations of the Hessian $\delta^2 f/\delta\hat{Q}^2$ for the obtained RS solution (see Section 3.6) demonstrate that this solution becomes *unstable* ($\det(\delta^2 f/\delta\hat{Q}^2) < 0$) in the entire low-temperature region $T < T_c$ [21]. It means that the true solution must be somewhere beyond the replica symmetric subspace.

3.3 Replica symmetry breaking

Because the RS solution appears to be unsatisfactory, we should try for the matrix $\hat{Q}$ with some other structure that contains more parameters. Within this new subspace we have to calculate the extremum of the replica free energy $f[\hat{Q}]$. After that, to check the stability of the obtained solution, we have to calculate the corresponding Hessian $\delta^2 f/\delta\hat{Q}^2$.

Actually, the situation appears to be much more sophisticated because (as we will see later) no ansatz that contains a finite number of parameters can provide a stable solution. Nevertheless, trying with different structures of $\hat{Q}$, and calculating the eigenvalues of the Hessian, one at least would be able to judge which ansatz could be better (that is to say, which is less unstable). Such a procedure could point the correct 'direction' in the space of the matrices $\hat{Q}$ towards the true solution.

The strategy of finding the true solution for the replica matrix $\hat{Q}$ in the limit $n \to 0$ is called the Parisi replica symmetry breaking (RSB) scheme [1, 22]. This is the infinite sequence of the ansatzs that approximate the true solution better and better. Eventually, the true solution can be formulated in terms of the continuous function, which is defined as the limit of the infinite sequence. Moreover, in this limit one is able to prove the stability of the obtained solution.

Consider now, step by step, the way in which the solution is approximated.

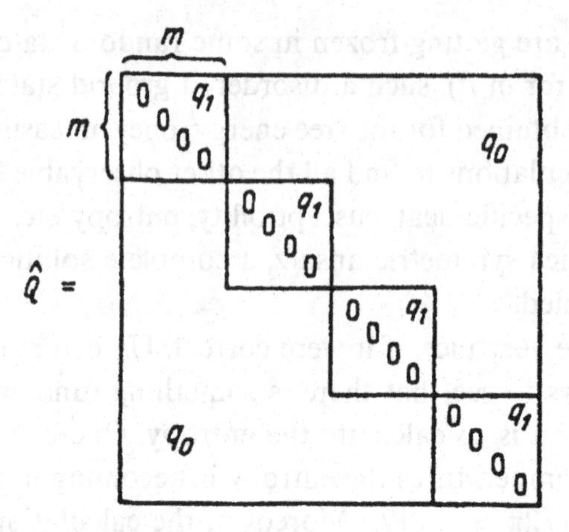

Fig. 3.1. The structure of the matrix Q_{ab} at the one-step replica symmetry breaking.

3.3.1 One-step RSB

At the first step, which is called the one-step RSB, it is 'natural' to divide n replicas into n/m groups, each containing m replicas (at this stage it is assumed that both m and n/m are integers). Then, the trial matrix $\hat{Q}$ is defined as follows: $Q_{ab} = q_1$, if the replicas a and b belong to the same group, and $Q_{ab} = q_0$, if the replicas a and b belong to different groups (the diagonal elements are equal to zero). In the compact form such a structure could be represented as follows:

$$Q_{ab} = \begin{cases} q_1 & \text{if } I\left(\dfrac{a}{m}\right) = I\left(\dfrac{b}{m}\right) \\ q_0 & \text{if } I\left(\dfrac{a}{m}\right) \neq I\left(\dfrac{b}{m}\right) \end{cases} \tag{3.24}$$

where $I(x)$ is the integer valued function, which is equal to the smallest integer larger than or equal to x. The qualitative structure of this matrix is shown in Fig. 3.1.

In the framework of the one-step RSB we have three parameters: q_1, q_2 and m, and these parameters have to be defined from the corresponding saddle-point equations. Using the explicit form of the matrix $\hat{Q}$ for the replica free energy (3.13), one gets:

$$f[\hat{Q}] = -\frac{1}{4}\beta + \frac{1}{2n}\beta \sum_{a<b}^{n} Q_{ab}^2 - \frac{1}{\beta n} \log\left(Z[\hat{Q}]\right) \tag{3.25}$$

where

$$Z[\hat{Q}] = \sum_{\sigma_a} \exp\left(\beta^2 \sum_{a<b}^{n} Q_{ab}\sigma_a\sigma_b\right) \tag{3.26}$$

Simple algebra yields:

$$\sum_{a<b}^{n} Q_{ab}\sigma_a\sigma_b = \frac{1}{2}\left[q_0\left(\sum_{a=1}^{n}\sigma_a\right)^2 + (q_1-q_0)\sum_{k=1}^{n/m}\left(\sum_{c_k=1}^{m}\sigma_{c_k}\right)^2 - nq_1\right] \quad (3.27)$$

Here k numbers the replica groups and c_k numbers the replicas inside the groups. After the Gaussian transformation in $Z[\hat{Q}]$ for each of the squares in the above equation, one gets:

$$\begin{aligned} Z[q_1,q_0,m] &= \int \frac{dz}{\sqrt{2\pi q_0}}\exp\left(-\frac{z^2}{2q_0}\right) \\ &\times \prod_{k=0}^{n/m}\left[\int \frac{dy_k}{\sqrt{2\pi(q_1-q_0)}}\exp\left(-\frac{y_k^2}{2(q_1-q_0)}\right)\right] \\ &\times \sum_{\sigma_a}\exp\left(\beta z\sum_{a}^{n}\sigma_a + \beta\sum_{k=0}^{n/m} y_k\left(\sum_{c_k=1}^{m}\sigma_{c_k}\right) - \frac{1}{2}\beta^2 nq_1\right) \end{aligned} \quad (3.28)$$

The summation over spins yields:

$$\begin{aligned} Z[q_1,q_0,m] &= \exp\left(-\frac{1}{2}\beta^2 nq_1\right)\int \frac{dz}{\sqrt{2\pi q_0}}\exp\left(-\frac{z^2}{2q_0}\right) \\ &\times \left[\int \frac{dy}{\sqrt{2\pi(q_1-q_0)}}\exp\left(-\frac{y^2}{2(q_1-q_0)}\right)[2\cosh\beta(z+y)]^m\right]^{n/m} \end{aligned} \quad (3.29)$$

For the second term in Eq. (3.25) one obtains:

$$\begin{aligned} \frac{\beta}{2n}\sum_{a<b}^{n} Q_{ab}^2 &= \frac{\beta}{4n}\left[q_1^2 m(m-1)\frac{n}{m} + q_0^2(n^2 - m^2\frac{n}{m})\right] \\ &= \frac{1}{4}\beta\left[q_1^2(m-1) + q_0^2(n-m)\right] \end{aligned} \quad (3.30)$$

Now the limit $n \to 0$ has to be taken. Originally the parameter m has been defined as an integer in the interval $1 \le m \le n$. The formal analytic continuation $n \to 0$ turns this interval into $0 \le m \le 1$, where m becomes a *continuous* parameter. Thus, taking the limit $n \to 0$ in Eqs. (3.29) (here the procedure of taking limit $n \to 0$ is similar to that described in Eq. (3.20) and (3.30) for the free energy (Eq. (3.25)) one obtains:

$$\begin{aligned} f(q_1,q_0,m) &= -\frac{1}{4}\beta\left[1 + mq_0^2 + (1-m)q_1^2 - 2q_1\right] - \frac{1}{m\beta}\int \frac{dz}{\sqrt{2\pi q_0}}\exp\left(-\frac{z^2}{2q_0}\right) \\ &\times \ln\left[\int \frac{dy}{\sqrt{2\pi(q_1-q_0)}}\exp\left(-\frac{y^2}{2(q_1-q_0)}\right)(2\cosh\beta(z+y))^m\right] \end{aligned} \quad (3.31)$$

One can easily check that in the extreme cases $m = 0$ and $m = 1$ the replica symmetric solution is recovered with $q = q_0$ and $q = q_1$ correspondingly.

It should be noted that in the framework of the RSB formalism one has to look for the *maximum* and not the minimum of the free energy. The formal reason is that in the limit $n \to 0$ the number of the components of the order parameter $\hat{Q}$ becomes *negative*. For example, in the case of the one-step RSB, each line of the matrix $\hat{Q}$ contains $(m-1) < 0$ components that are equal to q_1, and $(n-m) \to -m < 0$ components which are equal to q_0. This phenomenon can also be easily demonstrated for the case when the replica free energy (3.25) contains only the trivial term $(\beta/2n)\sum_{a<b} Q_{ab}^2$:

$$\lim_{n\to 0}\left[\frac{1}{2n}\beta\sum_{a<b} Q_{ab}^2\right] = -\frac{1}{4}\beta\left[(1-m)q_1^2 + mq_0^2\right] \tag{3.32}$$

Apparently, the 'correct extremum' of this free energy (in which the Hessian is positive) for $0 \leq m \leq 1$ is the maximum and not the minimum with respect to q_0 and q_1.

To derive the saddle-point equations for the parameters q_0, q_1 and m one just has to take the corresponding derivatives of the free energy (3.31). The calculations are straightforward, but as the resulting equations are rather cumbersome, we omit this simple exercise. The results of the numerical solution of these saddle-point equations are as follows.

(1) At $T < T_c = 1$ the function $f[q_1, q_0, m]$ indeed has the maximum at the non-trivial point: $0 < m(T) < 1$; $0 < q_0(T) < 1$; $0 < q_1(T) < 1$ (both for $T \to 1$ and $T \to 0$ one gets $m(T) \to 0$).

(2) Although at low temperatures the entropy of this solution becomes negative again, its absolute value appears to be much smaller than that of the RS solution: $S(T=0) \simeq -0.01$ (while for the RS solution $S(T=0) \simeq -0.17$).

(3) The most negative eigenvalue of the Hessian near T_c is equal to $-c(T-T_c)^2/9$ (c is some positive number), while for the RS solution it is equal to $-c(T-T_c)^2$. This could be interpreted as the instability of the solution being reduced by a factor of 9.

Thus, although the one-step RSB solution considered here turned out to be unsatisfactory too, it has appeared to be a much better approximation than the RS one. Therefore one could try to move further in the chosen 'direction' in the replica space.

Fig. 3.2. The grouping of replicas at the two-step replica symmmetry breaking.

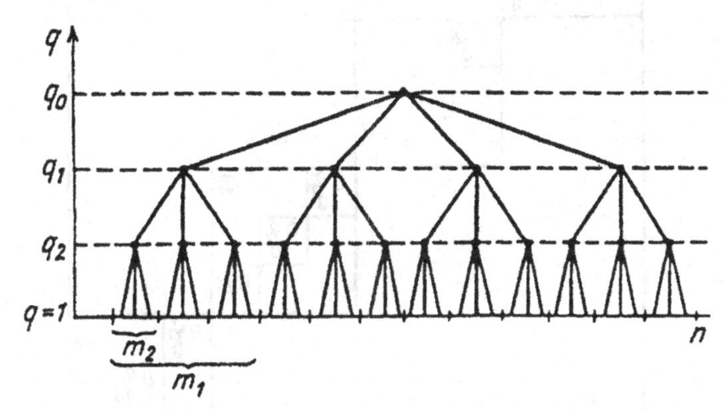

Fig. 3.3. The tree-like definition of the matrix elements Q_{ab} for the two-step RSB.

3.3.2 Full-scale RSB

Let us try to generalize the structure of the matrix $\hat{Q}$ for more steps of the replica symmetry breaking. Let us introduce a series of integers: $\{m_i\}$ $(i = 1, 2, \ldots, k+1)$ such that $m_0 = n, m_{k+1} = 1$ and all m_i/m_{i+1} at this stage are integers. Next, let us divide n replicas into n/m_1 groups such that each group would consist of m_1 replicas; each group of m_1 replicas divides into m_1/m_2 subgroups, so that each subgroup would consist of m_2 replicas; and so on (Fig. 3.2). Finally, the off-diagonal elements of the matrix $\hat{Q}$ let us define the following:

$$Q_{ab} = q_i, \text{ for } I\left(\frac{a}{m_i}\right) \neq I\left(\frac{b}{m_i}\right) \text{ and } I\left(\frac{a}{m_{i+1}}\right) = I\left(\frac{b}{m_{i+1}}\right); \ (i = 1, 2, \ldots, k+1) \tag{3.33}$$

where $\{q_i\}$ are a set of $(k+1)$ parameters ($k = 1$ corresponds to the case of the one-step RSB).

The above definition of the matrix elements can also be represented in terms of the hierarchical tree shown in Fig. 3.3: a particular matrix element Q_{ab} is equal to q_i corresponding to the level i of the tree, at which the lines radiating from the points a and b meet. The structure of the matrix $\hat{Q}$ for the case $k = 2$ is shown in Fig. 3.4.

Now we have to calculate the free energy, Eqs. (3.25)–(3.26), which depends on $(k+1)$ parameters q_i and k parameters m_i. After that, the limit $n \to 0$ has to be taken. Until the parameter n is an integer, according to the

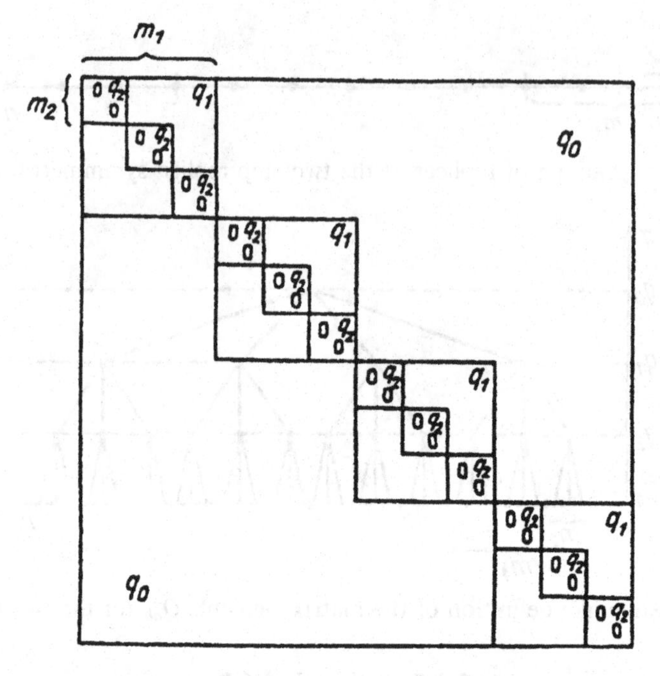

Fig. 3.4. The explicit form of the matrix Q_{ab} for the two-step RSB.

above definition of the $n \times n$ matrix $\hat{Q}$ the parameters $\{m_i\}$ must satisfy the inequalities $1 \leq m_{i+1} \leq m_i \leq n$. After the analytic continuation to the limit $n \to 0$ these inequalities turn into $0 \leq m_i \leq m_{i+1} \leq 1$.

The calculation of the free energy is similar to that of the one-step RSB case. After somewhat painful algebra the result obtained for the limit $n \to 0$ is in the following:

$$
\begin{aligned}
f\,[q_0, q_1, \ldots, q_k; m_1, m_2, \ldots, m_k] &= -\frac{1}{4}\beta \left[1 + \sum_{i=1}^{k}(m_{i+1} - m_i)q_i^2 - 2q_k\right] \\
&- \frac{1}{m_1}\int \frac{dz_0}{\sqrt{2\pi q_0}} \times \exp\left(-\frac{z_0^2}{2q_0}\right) \times \ln\left(\int dz_1 \frac{\exp\left(-\frac{z_1^2}{2(q_1-q_0)}\right)}{\sqrt{2\pi(q_1 - q_0)}}\right. \\
&\times \left[\int dz_2 \frac{\exp\left(-\frac{z_2^2}{2(q_2-q_1)}\right)}{\sqrt{2\pi(q_2 - q_1)}}\left[\ldots\left[\int dz_k \frac{\exp\left(-\frac{z_k^2}{2(q_k-q_{k-1})}\right)}{\sqrt{2\pi(q_k - q_{k-1})}}\right.\right.\right. \\
&\left.\left.\left.\times \left(2\cosh\beta\left(\sum_{i=0}^{k} z_k\right)\right)^{m_k}\right]^{m_{k-1}/m_k} \ldots\right]^{m_2/m_3}\right]^{m_1/m_2}\Bigg)
\end{aligned}
\tag{3.34}
$$

Finally, the parameters q_i and m_i have to be obtained from the saddle-point equations:

$$\frac{\partial f}{\partial q_i} = 0; \quad \frac{\partial f}{\partial m_i} = 0 \tag{3.35}$$

Unfortunately, it is hardly possible to obtain the explicit analytic solutions of these equations for an arbitrary k. Nevertheless, for a given (not very large) value of k these equations can be solved numerically and, in particular, for $k = 3$ the numerical solution for the zero temperature entropy gives the result $S(T = 0) \simeq -0.003$. In general, one finds that the more steps of the RSB are taken the less unstable the corresponding solution is. This indicates that presumably the true stable solution could be found in the limit $k \to \infty$. In this limit the infinite set of parameters q_i can be described in terms of the order parameter *function* $q(x)$ defined in the interval $(0 \le x \le 1)$. This function is obtained from the discrete step-like function:

$$q(x) = q_i, \quad \text{for } 0 \le m_i < x < m_{i+1} \le 1; \quad (i = 0, 1, \ldots, k) \tag{3.36}$$

in the limit of infinite number of steps, $k \to \infty$. In these terms the free energy becomes the functional of the function $q(x)$, and then the problem can be formulated as the searching for the maximum of this functional with respect to all (physically sensible) functions $q(x)$:

$$\frac{\delta f}{\delta q(x)} = 0 \tag{3.37}$$

For an arbitrary temperature $T < T_c$ the solution of this equation can be found only numerically. Nevertheless, near T_c all the calculations can be performed analytically, and the order parameter function $q(x)$ can be found explicitly (see Section 3.5 below). This solution appears to be quite helpful for attaining a qualitative physical understanding of what is going on in the low-temperature spin-glass phase (see Chapter 4). However, before proceeding with these calculations it is necessary to stop for a brief review of the formal general properties of the Parisi RSB matrices, which will be widely used in the further considerations.

3.4 Parisi RSB algebra

Using the definitions of the previous section one can easily prove that the linear space of the Parisi matrices, when completed with the identity $I_{ab} = \delta_{ab}$, is closed with respect to the matrix product $(QP)_{ab} = \sum_c Q_{ac}P_{cb}$ and the Hadamard product $(Q \cdot P)_{ab} = Q_{ab}P_{ab}$; by means of this operation it is possible to build polynomials that are invariant by permutations of replica indices.

Consider a generic Parisi matrix $\hat{Q}$, which in the *continuum limit* $k \to \infty$ for an *arbitrary value* of the parameter $n < 1$ is parametrized by its diagonal element $\tilde{q}$ and the off-diagonal function $q(x)$ $(n \leq x \leq 1)$: $\hat{Q} \to (\tilde{q}, q(x))$. Then for the linear invariants $\text{Tr}\hat{Q}$ and $\sum_{ab} Q_{ab}$ one can easily prove:

$$\text{Tr}\hat{Q} = n\tilde{q} \tag{3.38}$$

and

$$\lim_{k\to\infty} \sum_{a,b}^{n} Q_{ab} = n\tilde{q} + \lim_{k\to\infty} \left[n \sum_{i=0}^{k} (m_i - m_{i+1}) q_i \right] = n\tilde{q} - n \int_n^1 dx q(x) \tag{3.39}$$

Similarly to the above equation one gets:

$$\lim_{k\to\infty} \sum_{a,b}^{n} Q_{ab}^{l} = n\tilde{q}^{l} - n \int_n^1 dx q^{l}(x) \tag{3.40}$$

where the power l can be arbitrary.

Now let A and B be two Parisi matrices parametrized respectively by $(\tilde{a}, a(x))$ and $(\tilde{b}, b(x))$. Then for an arbitrary finite n for the Hadamard product $(Q \cdot P)_{ab} = Q_{ab} P_{ab}$ one easily proves:

$$A \cdot B \to \left(\tilde{a}\tilde{b},\ a(x) b(x) \right) \tag{3.41}$$

Let us denote the parametrization of the matrix product of the two matrices as follows: $AB \to (\tilde{c}, c(x))$. Then after somewhat painful algebra one can prove that

$$\begin{aligned} \tilde{c} &= \tilde{a}\tilde{b} - \langle ab \rangle \\ c(x) &= -n a(x) b(x) + [\tilde{a} - \langle a \rangle] b(x) + [\tilde{b} - \langle b \rangle] a(x) \\ &\quad - \int_n^x dy [a(x) - a(y)][b(x) - b(y)] \end{aligned} \tag{3.42}$$

where we have introduced the notation:

$$\langle a \rangle \equiv \int_n^1 dx a(x) \tag{3.43}$$

For the eigenvalues of a Parisi matrix Q and their multiplicities one finds:

$$\begin{aligned} \lambda_0 &= \tilde{a} - \langle a \rangle; & &\text{with multiplicity } 1 \\ \lambda(x) &= \tilde{a} - x a(x) - \int_x^1 dy q(y); & &\text{with multiplicity } -\frac{n}{x^2} dx \end{aligned} \tag{3.44}$$

where $x \in [n, 1]$. The above algorithms are sufficient to operate quite easily with the Parisi matrices in the continuum RSB representation.

3.5 RSB solution near T_c

Near the critical temperature $T_c = 1$, the solution for the saddle-point function $q(x)$ can be obtained analytically. In the vicinity of the phase transition point the order parameter $q(x)$ should be expected to be small in $\tau = (T_c - T)/T_c \ll 1$, and consequently one can expand the replica free energy (3.25)–(3.26) in powers of the matrix Q_{ab}. This calculation is straightforward, and the result of the expansion up to the fourth order is as follows:

$$f[\hat{Q}] = \lim_{n\to 0}\frac{1}{n}\left[-\frac{1}{2}\tau \mathrm{Tr}(\hat{Q})^2 - \frac{1}{6}\mathrm{Tr}(\hat{Q})^3 - \frac{1}{12}\sum_{a,b} Q_{ab}^4 + \frac{1}{4}\sum_{a,b,c} Q_{ab}^2 Q_{ac}^2 - \frac{1}{8}\mathrm{Tr}(\hat{Q})^4\right] \tag{3.45}$$

Here in all the terms but the first one we have substituted $T = 1$.

Detailed study of the stability of the replica symmetric solution shows that it is the term $\sum_{a,b} Q_{ab}^4$ that makes the RS solution unstable below T_c, and it is this term which is responsible for the replica symmetry breaking [20]. This indicates that for the RSB solution near T_c, the last two terms of the fourth order in (3.45) must be of higher orders in τ than all the other terms. Thus, the easiest way to obtain the solution is by first neglecting these last two terms; and then by using the explicit form of the obtained solution for $q(x)$ one can easily prove *a posteriori* that these neglected terms are indeed of higher orders in τ.

Using the rules for the Parisi matrices in the continuum RSB representation described in the previous section one can easily get the explicit expression for the free energy as the functional of $q(x)$. In particular, using Eq. (3.42) for the second term in Eq. (3.45) after simple algebra in the limit $n \to 0$ one gets:

$$\lim_{n\to 0}\frac{1}{n}\mathrm{Tr}(\hat{Q})^3 = \int_0^1 dx\left[xq^3(x) + 3q(x)\int_0^x dyq^2(y)\right] \tag{3.46}$$

The first and the third terms in Eq. (3.45) can be expressed using Eq. (3.40) (in our case $\tilde{q} \equiv 0$). For the free energy one finally obtains:

$$f[q(x)] = \frac{1}{2}\int_0^1 dx\left[\tau q^2(x) - \frac{1}{3}xq^3(x) - q(x)\int_0^x dyq^2(y) + \frac{1}{6}q^4(x)\right] \tag{3.47}$$

Variation of this expression with respect to the function $q(x)$ yields the following saddle-point equation:

$$2\tau q(x) - xq^2(x) - 2q(x)\int_x^1 dyq(y) - \int_0^x dyq^2(y) + \frac{2}{3}q^3(x) = 0 \tag{3.48}$$

The solution of this equation is simple. Taking the derivative of Eq. (3.48) over x one gets:

$$q'(x)\left[2\tau - 2xq(x) - 2\int_x^1 dyq(y) + 2q^2(x)\right] = 0 \tag{3.49}$$

This equation results in the following:

$$2\tau - 2xq(x) - 2\int_x^1 dyq(y) + 2q^2(x) = 0 \tag{3.50}$$

or

$$q'(x) = 0 \tag{3.51}$$

The last equation means that $q(x) = \text{const}$, and it corresponds to the replica symmetric solution that has been already studied. Consider Eq. (3.50). Taking the derivative over x again, one gets:

$$q(x) = \frac{1}{2}x \tag{3.52}$$

The above simple analysis allows us to build an ansatz for a general form of the solution of the original saddle-point equation (3.48):

$$q(x) = \begin{cases} q_0, & 0 \le x \le x_0 \\ \frac{1}{2}x, & x_0 \le x \le x_1 \\ q_1, & x_1 \le x \le 1 \end{cases} \tag{3.53}$$

where

$$x_1 = 2q_1\ ; \qquad x_0 = 2q_0 \tag{3.54}$$

Substituting Eq. (3.53) into the original saddle-point equation (3.48) one obtains two equations for two unknown parameters q_0 and q_1:

$$\begin{aligned} q_0\left[2\tau - 2q_1 + 2q_1^2\right] - \frac{4}{3}q_0^3 = 0 \\ q_1\left[2\tau - 2q_1 + 2q_1^2\right] - \frac{4}{3}q_0^3 = 0 \end{aligned} \tag{3.55}$$

The solution of these equations is:

$$\begin{aligned} q_0 &= 0 \\ q_1 &= \tau + O(\tau^2) \end{aligned} \tag{3.56}$$

Now one can easily check that the last two terms of the fourth order in Eq. (3.45) are of a higher order in τ compared to the other terms. It appears that because they contain additional summations over replicas, in the continuum limit representation this results in additional integrations over x that eventually provide additional powers of τ.

Note that the obtained RSB solution could be easily generalized for the case of non-zero external magnetic field represented in the original Ising spin Hamiltonian (3.1) by the term $h\sum_i \sigma_i$. As a simple exercise one can easily derive that if the value of the field h is small in the corresponding expression for the functional RSB free energy (near T_c), (Eq. (3.47)), the magnetic field is represented by the additional term $h^2 q(x)$. This does not change the structure of the saddle-point solution (3.53), but in the r.h.s. of Eqs. (3.55) for the parameters q_0 and q_1 one gets h^2 instead of zero. Then, in the leading order in τ and h the value of q_1 does not change, while the parameter q_0 (and x_0) becomes non-zero:

$$q_0 \sim h^{2/3} \tag{3.57}$$

Thus, at the critical value of the field

$$h_c(\tau) \simeq \tau^{3/2} \tag{3.58}$$

(when $x_0 = x_1$ and $q_0 = q_1$) the solution for $q(x)$ becomes replica symmetric.

3.6 de Almeida–Thouless line

Actually, the equation for the critical line $h_c(T)$, defined by the condition that for $h > h_c(T)$ the replica symmetric solution is stable, can be obtained for the whole range of temperatures and magnetic fields [21]. This line is usually called the *de Almeida–Thouless* (AT) line.

Let us derive the criteria for the stability of the replica symmetric solution of the SK model in a non-zero external magnetic field. The general expression for the replica partition function of the SK model in the external field h is (see Section 3.2):

$$Z_n = \int D\hat{Q} \exp\left(-\beta n N f[\hat{Q}]\right) \tag{3.59}$$

where $f[\hat{Q}]$ is the replica free energy:

$$f[\hat{Q}] = -\frac{1}{4}\beta + \frac{\beta}{2n}\sum_{a<b}^{n} Q_{ab}^2 - \frac{1}{\beta n}\log\left[\sum_{\sigma_a}\exp\left(\beta^2\sum_{a<b}^{n} Q_{ab}\sigma_a\sigma_b + \beta h\sum_{a}^{n}\sigma_a\right)\right] \tag{3.60}$$

Here Q_{ab} is the symmetric $n \times n$ replica matrix with zero diagonal elements. The replica symmetric solution of this model is defined by the condition: $Q_{ab} = q$ for all $a \neq b$, where the value of $q(T,h)$ is given by the equation (see Eq. (3.22)):

$$q = \int_{-\infty}^{+\infty} \frac{dz}{\sqrt{2\pi}} \tanh^2(\beta\sqrt{q}z + \beta h)\exp\left(-\frac{1}{2}z^2\right) \tag{3.61}$$

To check whether the RS solution defines the correct extremum of the free energy (3.60) we have to consider small deviations from the RS form of the matrix Q_{ab}:

$$Q_{ab} = q + \eta_{ab} \tag{3.62}$$

where η_{ab} is also a symmetric $n \times n$ replica matrix with zero diagonal elements. The expansion of the free energy (3.60) up to the second order in η_{ab} is:

$$f[q, \hat{\eta}] = f_{\mathrm{rs}}(q) - \frac{\beta}{2n} \sum_{(ab),(cd)} G_{(ab)(cd)} \eta_{ab} \eta_{cd} \tag{3.63}$$

where $f_{\mathrm{rs}}(q)$ is the replica symmetric free energy, and

$$G_{(ab)(cd)} = \delta_{(ab)(cd)} - \beta^2 \left[\langle \sigma_a \sigma_b \sigma_c \sigma_d \rangle - \langle \sigma_a \sigma_b \rangle \langle \sigma_c \sigma_d \rangle \right] \tag{3.64}$$

The RS solution will be stable only if the quadratic form in Eq. (3.63) is positively defined. The matrix $\hat{G}$ has three different types of matrix element:

$$G_{(ab)(ab)} = 1 - \beta^2 (1 - \langle \sigma_a \sigma_b \rangle^2) \equiv A$$

$$G_{(ab)(ac)} = -\beta^2 \left[\langle \sigma_b \sigma_c \rangle - \langle \sigma_a \sigma_b \rangle^2 \right] \equiv B, \quad (b \neq c)$$

$$G_{(ab)(cd)} = -\beta^2 \left[\langle \sigma_a \sigma_b \sigma_c \sigma_d \rangle - \langle \sigma_a \sigma_b \rangle^2 \right] \equiv C, \quad ((a,b) \neq (c,d)) \tag{3.65}$$

Here $\langle \sigma_a \sigma_b \rangle = q$ and

$$\langle \sigma_a \sigma_b \sigma_c \sigma_d \rangle \equiv r = \int_{-\infty}^{+\infty} \frac{dz}{\sqrt{2\pi}} \tanh^4(\beta \sqrt{q} z + \beta h) \exp\left(-\frac{1}{2} z^2 \right) \tag{3.66}$$

Because the matrix $\hat{G}$ is real symmetric and its order is $\frac{1}{2} n(n-1)$, the number of linearly independent eigenvectors must also be equal to $\frac{1}{2} n(n-1)$. The complete set of these eigenvectors can be found for general values of n, and then the limit $n \to 0$ can be taken. Let us consider the corresponding eigenvalue equation:

$$\hat{G} \eta = \lambda \eta \tag{3.67}$$

One can easily check that the complete set of eigenvectors and eigenvalues defined by this equation can be classified in terms of three different types of symmetry.

(1) The eigenvectors $\eta^{(1)}$ that are symmetric under interchange of all indices:

$$\eta_{ab}^{(1)} = \eta \tag{3.68}$$

for all $a \neq b$. Substituting this into the eigenvalue equation (3.67) and taking into account Eq. (3.65), one gets:

$$\left[A + 2(n-2)B + \frac{1}{2}(n-2)(n-3)C - \lambda\right] \eta = 0 \tag{3.69}$$

This equation yields one non-degenerate eigenvalue:

$$\lambda_1 = A + 2(n-2)B + \frac{1}{2}(n-2)(n-3)C \tag{3.70}$$

(2) The eigenvectors $\eta^{(2)}$ that are symmetric under interchange of all but one of the indices (this fixed index could be taken e.g. number 1):

$$\begin{aligned} \eta_{ab}^{(2)} &= \eta_1^{(2)}; \quad \text{for } a \text{ or } b = 1 \\ \eta_{ab}^{(2)} &= \eta_2^{(2)}; \quad \text{for } a, b \neq 1 \end{aligned} \tag{3.71}$$

To ensure orthogonality with the eigenvector (3.68) one should impose the condition: $\sum_{ab} \eta_{ab}^{(2)} \eta_{ab}^{(1)} = 0$, which gives: $\eta_1^{(2)} = (1 - \frac{1}{2}n)\eta_2^{(2)}$. There exist $(n-1)$ such vectors. Substituting (3.71) into the eigenvalue equation (3.67) one obtains:

$$[A + (n-4)B - (n-3)C - \lambda]\, \eta_1^{(2)} = 0 \tag{3.72}$$

which yields the following eigenvalue:

$$\lambda_2 = A + (n-4)B - (n-3)C \tag{3.73}$$

with degeneracy $(n-1)$.

(3) The eigenvectors $\eta^{(3)}$ that are symmetric under interchange of all but two of the indices (these could be taken to be e.g. number 1 and number 2):

$$\begin{aligned} \eta_{12}^{(3)} &= \eta_1^{(3)}; \\ \eta_{1a}^{(3)} &= \eta_{2a}^{(3)} = \eta_2^{(3)}; \quad \text{for } a \neq 1, 2 \\ \eta_{ab}^{(3)} &= \eta_3^{(3)}; \qquad\qquad \text{for } a, b \neq 1, 2 \end{aligned} \tag{3.74}$$

Orthogonality with the previous eigenvectors, (3.68) and (3.71), imposes the conditions: $\sum_{ab} \eta_{ab}^{(1)} \eta_{ab}^{(3)} = 0$ and $\sum_{ab} \eta_{ab}^{(2)} \eta_{ab}^{(3)} = 0$, which yields:

$$\begin{aligned} \eta_1^{(3)} &= (2-n)\eta_2^{(3)}; \\ \eta_2^{(3)} &= \frac{1}{2}(3-n)\eta_3^{(3)} \end{aligned} \tag{3.75}$$

One can easily see that the number of such vectors is $\frac{1}{2}n(n-3)$.

Substituting (3.74) and (3.75) into the eigenvalue equation (3.67) one obtains the following eigenvalue:

$$\lambda_3 = A - 2B + C \tag{3.76}$$

which has the degeneracy $\frac{1}{2}n(n-3)$.

In the limit $n \to 0$ one gets:

$$\begin{aligned} \lambda_1 = \lambda_2 &= A - 4B + 3C \\ \lambda_3 &= A - 2B + C \end{aligned} \tag{3.77}$$

Substituting the values of A, B and C, Eqs. (3.65)–(3.66) into the condition that the first two eigenvalues in (3.77) are positive (which ensures the stability of the RS solution) one finds:

$$\int_{-\infty}^{+\infty} \frac{dz}{\sqrt{2\pi}} \left[1 - 4\tanh^2(\beta\sqrt{q}z + \beta h) + 3\tanh^4(\beta\sqrt{q}z + \beta h)\right] \exp\left(-\frac{1}{2}z^2\right) < T^2 \tag{3.78}$$

where q is defined by the Eq. (3.61). It can be easily shown that this condition is satisfied for any values of T and h. Therefore, the stability of the RS solution is controlled by the condition $\lambda_3 > 0$, which after simple algebra can be represented as follows:

$$\int_{-\infty}^{+\infty} \frac{dz}{\sqrt{2\pi}} \cosh^{-4}(\beta z\sqrt{q} + \beta h) \exp\left(-\frac{1}{2}z^2\right) < T^2 \tag{3.79}$$

Thus, the line of stability of the RS solution in the (T, h) plane, the AT-line, is defined by the following two equations:

$$\begin{aligned} \int_{-\infty}^{+\infty} \frac{dz}{\sqrt{2\pi}} \cosh^{-4}(\beta z\sqrt{q} + \beta h) \exp\left(-\frac{1}{2}z^2\right) &= T^2 \\ q = \int_{-\infty}^{+\infty} \frac{dz}{\sqrt{2\pi}} \tanh^2(\beta z\sqrt{q} + \beta h) \exp\left(-\frac{1}{2}z^2\right) \end{aligned} \tag{3.80}$$

It can easily be shown that in the limit of small fields, $h \to 0$, the AT-line comes to the phase transition temperature $T_c = 1$. Expanding Eqs. (3.80) in powers of small q and h one gets the following stability condition (see Eq. (3.58)):

$$\left(1 - \frac{T}{T_c}\right) < \left(\frac{3}{4}\right)^{1/3} h^{2/3} \tag{3.81}$$

where $(1 - T/T_c) \ll 1$. It can also easily be shown that in the opposite limit of large fields, $h \gg 1$, the corresponding temperature

$T_{\mathrm{AT}}(h)$ appears to be exponentially small (in this case the value of q is close to 1):

$$T > T_{\mathrm{AT}}(h) = \frac{4}{3\sqrt{2\pi}} \exp\left(-\frac{1}{2}h^2\right) \tag{3.82}$$

4

Physics of replica symmetry breaking

In this chapter the physical interpretation of the formal RSB solution will be proposed, and some new concepts and quantities will be introduced. The crucial concept that is needed to understand physics behind the RSB structures is that of the *pure states.*

4.1 The pure states

Consider again a simple example of the ferromagnetic system. Here, spontaneous symmetry breaking takes place below the critical temperature T_c, and at each site the non-zero spin magnetizations $\langle\sigma_i\rangle = \pm m$ appear. As we have already discussed in Section 2.2, in the thermodynamic limit the two ground states with the global magnetizations $\langle\sigma_i\rangle = +m$ and $\langle\sigma_i\rangle = -m$ are separated by an infinite energy barrier. Therefore, once the system has happened to be in one of these states, it will never be able (during any finite time) to jump into the other one. In this sense, the *observable* state is not the Gibbs one (which is obtained by summing over all the states), but one of these two states with non-zero global magnetizations. To distinguish them from the Gibbs state they could be called the 'pure states'. More formally, the pure states could also be defined by the property that all the connected correlation functions in these states, such as $\langle\sigma_i\sigma_j\rangle_c \equiv \langle\sigma_i\sigma_j\rangle - \langle\sigma_i\rangle\langle\sigma_j\rangle$, tend towards zero at large distances.

In the previous chapter we obtained a special type of spin-glass ground-state solution. Formally this solution is characterized by the RSB in the corresponding order parameter matrix Q_{ab}, and it automatically means that there actually exist many other solutions of this type. This fact is a direct consequence of the symmetry of the replica free energy (Eqs. (3.25)–(3.26)) with respect to permutations of replicas: if there exists a particular solution for the matrix $\hat{Q}_*$ with the RSB, then any other matrix obtained via permutations of the replica indices in $\hat{Q}_*$ will also be a solution. On the other hand, as the total mean-field free energy (which is the function of $\hat{Q}$) is proportional to the volume of the system, the energy barriers separating the corresponding ground states must be infinite in the thermodynamic limit. Consequently, just like in the example of the ferromagnetic system, all these RSB states could be called the pure states of the low-temperature spin-glass phase. Correspondingly, the Gibbs state (which is formally obtained by summing over *all* the states of the system) could be considered as being given by the summation over all the pure states with the corresponding thermodynamic weight defined by values of their free energies.

For instance, the thermodynamic (Gibbs) average of the site magnetizations can be represented as follows:

$$\langle\sigma_i\rangle \equiv m_i = \sum_{\alpha} w_\alpha m_i^\alpha \tag{4.1}$$

Here m_i^α are the site magnetizations in the pure state α, and w_α denotes its statistical weight, which formally can be represented as follows:

$$w_\alpha = \exp(-F_\alpha) \tag{4.2}$$

where F_α is the free energy corresponding to this pure state. In the same way the two-point correlation function can be represented as the linear combination

$$\langle\sigma_1\sigma_2\rangle = \sum_{\alpha} w_\alpha \langle\sigma_1\sigma_2\rangle_\alpha \tag{4.3}$$

where $\langle\sigma_1\sigma_2\rangle_\alpha$ is the two-point correlation function in the pure state number α. According to the definition of the pure state

$$\langle\sigma_1\sigma_2\rangle_\alpha = \langle\sigma_1\rangle_\alpha\langle\sigma_2\rangle_\alpha \tag{4.4}$$

Similar expressions can be written for any many-point correlation functions.

The representation of the thermodynamic Gibbs state as a linear combination of the pure states in which all extensive quantities have vanishing long-distance fluctuations, is actually a central point in the exact definition of the concept of the spontaneous symmetry breaking in statistical mechanics.

4.2 Physical order parameter $P(q)$ and the replica solution

To investigate the statistical properties of the spin-glass pure states let us define the *overlaps* $\{q_{\alpha\beta}\}$ among them as follows:

$$q_{\alpha\beta} \equiv \frac{1}{N}\sum_{i}^{N} m_i^\alpha m_i^\beta \tag{4.5}$$

where $m_i^\alpha = \langle\sigma_i\rangle_\alpha$ and $m_i^\beta = \langle\sigma_i\rangle_\beta$ are the site magnetizations in the pure states α and β. Apparently, $0 \leq | q_{\alpha\beta} | \leq 1$. To describe the statistics of these overlaps it is natural to introduce the following probability distribution function:

$$P_J(q) = \sum_{\alpha\beta} w_\alpha w_\beta \delta(q_{\alpha\beta} - q) \tag{4.6}$$

Note that this distribution function is defined for a given sample, and it can depend on a concrete realization of the quenched interactions J_{ij}. The 'observable' distribution function should, of course, be averaged over the disorder parameters:

$$P(q) = \overline{P_J(q)} \tag{4.7}$$

The distribution function $P(q)$ gives the probability of finding two pure states having their overlap equal to q, with the condition that these states are taken with their statistical thermodynamic weights $\{w_\alpha\}$.

It is the distribution function $P(q)$ that can be considered as the physical order parameter. It should be stressed that $P(q)$ is a much more general concept than ordinary order parameters, which usually describe the phase transitions in ordinary statistical systems. The fact that it is a *function* is actually a manifestation of the crucial phenomenon that for the description of the spin-glass phase one needs an infinite number of order parameters. The non-trivial structure of this distribution function (it will be calculated explicitly below) demonstrates that the properties of the spin-glass state are essentially different from those of traditional magnets.

Consider now which way the order parameter function $P(q)$ can be calculated in terms of the replica method. Let us introduce the following set of correlation functions:

$$\begin{aligned} q_J^{(1)} &= \frac{1}{N}\sum_i \langle\sigma_i\rangle^2 \\ q_J^{(2)} &= \frac{1}{N^2}\sum_{i_1 i_2} \langle\sigma_{i_1}\sigma_{i_2}\rangle^2 \\ &\dots \\ q_J^{(k)} &= \frac{1}{N^k}\sum_{i_1 \dots i_k} \langle\sigma_{i_1}\dots\sigma_{i_k}\rangle^2 \end{aligned} \tag{4.8}$$

Using the representation of the Gibbs averages in terms of the pure states (4.3)–(4.4) for the correlation functions (4.8) one gets:

$$q_J^{(1)} = \frac{1}{N}\sum_i \left(\sum_\alpha w_\alpha \langle\sigma_i\rangle_\alpha\right)\left(\sum_\beta w_\beta \langle\sigma_i\rangle_\beta\right)$$

$$
\begin{aligned}
&= \sum_{\alpha\beta} w_\alpha w_\beta \left(\frac{1}{N} \sum_i \langle\sigma_i\rangle_\alpha \langle\sigma_i\rangle_\beta \right) \\
&= \sum_{\alpha\beta} w_\alpha w_\beta q_{\alpha\beta} \\
&= \int dq P_J(q) q; \\
q_J^{(2)} &= \frac{1}{N^2} \sum_{i_1 i_2} \left(\sum_\alpha w_\alpha \langle\sigma_{i_1}\sigma_{i_2}\rangle_\alpha \right) \left(\sum_\beta w_\beta \langle\sigma_{i_1}\sigma_{i_2}\rangle_\beta \right) \\
&= \sum_{\alpha\beta} w_\alpha w_\beta \left(\frac{1}{N} \sum_{i_1} \langle\sigma_{i_1}\rangle_\alpha \langle\sigma_{i_1}\rangle_\beta \right) \left(\frac{1}{N} \sum_{i_2} \langle\sigma_{i_2}\rangle_\alpha \langle\sigma_{i_2}\rangle_\beta \right) \\
&= \sum_{\alpha\beta} w_\alpha w_\beta (q_{\alpha\beta})^2 \\
&= \int dq P_J(q) q^2; \\
&\dots \\
q_J^{(k)} &= \int dq P_J(q) q^k
\end{aligned}
\tag{4.9}
$$

For the corresponding correlation functions averaged over the disorder from Eqs. (4.8)–(4.9) one gets:

$$
\begin{aligned}
q^{(1)} &\equiv \overline{q_J^{(1)}} = \overline{\langle\sigma_i\rangle^2} = \int dq P(q) q \\
&\dots \\
q^{(k)} &\equiv \overline{q_J^{(k)}} = \overline{\langle\sigma_{i_1} \dots \sigma_{i_k}\rangle^2} = \int dq P(q) q^k
\end{aligned}
\tag{4.10}
$$

where $i_1 \neq i_2 \neq \dots \neq i_k$.

The crucial point in the above consideration is that the function $P(q)$ originally defined to describe the statistics of (somewhat abstract) pure states, can be calculated (at least theoretically) from the multipoint correlation functions in the *Gibbs states.* Therefore, if we were able to calculate the above multipoint correlation functions in terms of the replica approach, the connection of the formal RSB scheme with the physical order parameter would be established.

In terms of the replica approach, the correlator $q^{(1)} = \overline{\langle\sigma_i\rangle^2}$ can be represented as follows:

$$
q^{(1)} = \overline{\frac{1}{Z^2} \sum_\sigma \sum_s (\sigma_i s_i) \exp(-\beta H[\sigma] - \beta H[s])}
$$

$$= \lim_{n\to 0}\left[\prod_{a=1}^{n}\sum_{\sigma^a}\right](\sigma_i^b\sigma_i^c)\exp\left(-\beta\sum_{a=1}^{n}H[\sigma^a]\right)$$

$$\equiv \lim_{n\to 0}\overline{\langle\sigma_i^b\sigma_i^c\rangle} \qquad (b\neq c) \tag{4.11}$$

where a and b are two fixed different replicas (the summation over the remaining $(n-2)$ replicas in Eq. (4.11) gives the factor Z^{n-2}, which turns into Z^{-2} in the limit $n\to 0$). In a similar way one gets:

$$q^{(2)} = \lim_{n\to 0}\overline{\langle\sigma_{i_1}^a\sigma_{i_2}^a\sigma_{i_1}^b\sigma_{i_2}^b\rangle} \qquad (i_1\neq i_2;\ a\neq b)$$

$$\dots$$

$$q^{(k)} = \lim_{n\to 0}\overline{\langle\sigma_{i_1}^a\dots\sigma_{i_k}^a\sigma_{i_1}^b\dots\sigma_{i_k}^b\rangle}; \quad (i_1\neq i_2\neq\dots\neq i_k;\ a\neq b) \tag{4.12}$$

In the calculations of the previous chapter it has been demonstrated that the free energy of the model under consideration is factorizing into the independent site replica free energies. Therefore, the result (4.12) for $q^{(k)}$ can be represented as follows:

$$q^{(k)} = \lim_{n\to 0}\left[\,\overline{\langle\sigma_i^a\sigma_i^b\rangle}\,\right]^k = \lim_{n\to 0}[Q_{ab}]^k \tag{4.13}$$

where (see Eq. (3.9))

$$Q_{ab} = \overline{\langle\sigma_i^a\sigma_i^b\rangle} \tag{4.14}$$

is the replica order parameter matrix introduced in Chapter 3, which is obtained from the saddle-point equation for the replica free energy. Because in the RSB solution the matrix elements of Q_{ab} are not equivalent, one has to sum over *all* the saddle-point solutions for the matrix Q_{ab} to perform the Gibbs average. Such solutions can be obtained from one of the RSB solutions by doing all possible permutations of rows and columns in Q_{ab}. The summation over all these permutations corresponds to the summation over the replica subscripts a and b of the matrix Q_{ab}. Thus, the final result for the correlator $q^{(k)}$ should be represented as follows:

$$q^{(k)} = \lim_{n\to 0}\frac{1}{n(n-1)}\sum_{a\neq b}[Q_{ab}]^k \tag{4.15}$$

where $n(n-1)$ is the normalization factor that is equal to the number of different replica permutations.

The results (4.15) and (4.10) demonstrate that by using the formal RSB solution for the matrix Q_{ab} considered in the previous chapter one can calculate the order parameter distribution function $P(q)$, which has been originally introduced on the basis of purely qualitative physical arguments.

From these two equations one gets the following explicit expression for the distribution function $P(q)$:

$$P(q) = \lim_{n\to 0} \frac{1}{n(n-1)} \sum_{a\neq b} \delta(Q_{ab} - q) \tag{4.16}$$

Using the algorithms of the Parisi algebra (Section 3.4) in the continuous RSB representation this result can be rewritten as follows:

$$P(q) = \int_0^1 dx \delta(q(x) - q) \tag{4.17}$$

Assuming that the function $q(x)$ is monotonous (which is the case for the saddle-point solution obtained in Chapter 3), one can introduce the inverse function $x(q)$, and then from Eq. (4.17) one finally obtains:

$$P(q) = \frac{dx(q)}{dq} \tag{4.18}$$

(Note that the same result can be obtained by comparing Eqs. (4.15) and (4.10)). This is a key result, which defines the physical order parameter distribution function $P(q)$ in terms of the formal saddle-point Parisi function $q(x)$.

The above result can also be represented in the integral form:

$$x(q) = \int_0^q dq' P(q') \tag{4.19}$$

which gives the answer to the question, as to the physical meaning of the Parisi function $q(x)$. According to Eq. (4.19) the answer is as follows: the function $x(q)$ inverse to $q(x)$ gives the probability of finding a pair of the pure states that would have an overlap not bigger than q.

Using the explicit solution for the Parisi function $q(x)$ in the vicinity of the critical point (Eqs. (3.53)–(3.57)), according to Eq. (4.18) for the distribution function $P(q)$, one gets:

$$P(q) = x_0 \delta(q - q_0) + (1 - x_1)\delta(q - q_1) + p(q) \tag{4.20}$$

where $p(q)$ is the smooth function defined in the interval $q_0 \leq q \leq q_1$. In the vicinity of the critical point, $\tau \ll 1$, where the solution (3.53) is valid, this function is just constant: $p(q) = 2$.

The result (4.20) shows that the statistics of the overlaps of the pure states demonstrate the following properties:

(1) There is a finite probability $(1 - x_1) \simeq (1 - 2\tau)$ that, taken at random, two pure states would appear to be the same state. The 'selfoverlaps' (Eq. (2.3)), of these states is equal to $q_1 \simeq \tau$.

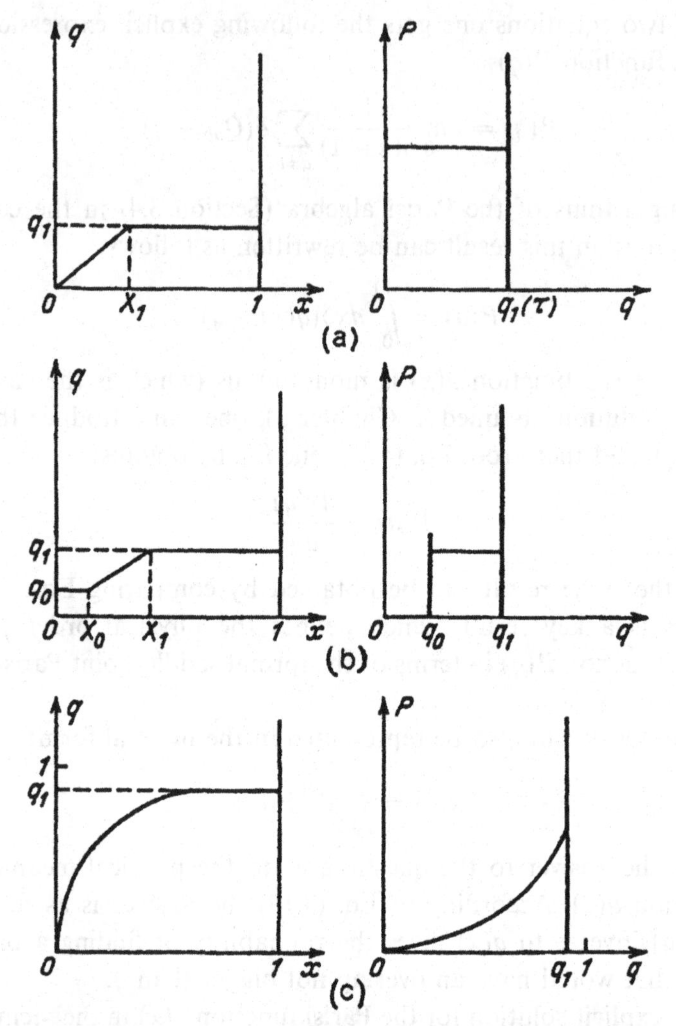

Fig. 4.1. The qualitative shape of the functions $q(x)$ and $P(q)$: (a) in the zero magnetic field near the critical point ($\tau \ll 1$); (b) in the finite magnetic field h, for $0 < h < h_c(T)$ and $\tau \ll 1$; (c) in the zero magnetic field and in the limit of low temperatures, $T \ll 1$.

(2) In the presence of non-zero external magnetic field h there is a finite probability $x_0 \sim h^{2/3}$ that taken at random two pure states would appear to be the most 'distant', having the minimum possible overlap $q_0 \sim h^{2/3}$.

(3) There is a finite probability $(x_1 - x_0)$ that, taken at random, two pure states would have the overlap q in the interval $q_0 \leq q \leq q_1$. For a given small interval δq there is a finite probability $p(q)\delta q$ of finding

two pure states with the overlap in the interval $(q, q + \delta q)$, where $q_0 \leq q \leq q_1$.

Although for arbitrary values of the temperature and the magnetic field it is hardly possible to calculate the functions $q(x)$ and $P(q)$ analytically, their qualitative behavior remains similar to the case considered above. The only difference is that the concrete shape of the function $P(q)$ in the interval $q_0 \leq q \leq q_1$ (as well as the function $q(x)$ at the interval $x_0 \leq x \leq x_1$) becomes less trivial. Also, the dependencies of x_0, x_1, q_0 and q_1 on the temperature and the magnetic field become more complicated.

The qualitative behavior of the functions $q(x)$ and $P(q)$ for different values of the temperature and the magnetic field are shown in Fig. 4.1.

5

Ultrametricity

5.1 Ultrametric structure of pure states

The solutions for the functions $q(x)$ and $P(q)$, obtained in the previous chapters, indicate that the structure of the space of the spin-glass pure states must be highly non-trivial. However, the distribution function $P(q)$ of the pure states overlaps does not give enough information about this structure. To get insight into the topology of the space of the pure states one needs to know the properties of the higher order correlations of the overlaps.

Let us consider the distribution function $P(q_1, q_2, q_3)$, which describes the joint statistics of the overlaps of *three* arbitrary pure states. By definition, for three arbitrary pure states α, β and γ this function gives the probability that their mutual overlaps $q_{\alpha\beta}, q_{\alpha\gamma}$ and $q_{\beta\gamma}$ are equal, corresponding to q_1, q_2 and q_3:

$$P(q_1, q_2, q_3) = \overline{\sum_{\alpha\beta\gamma} w_\alpha w_\beta w_\gamma \delta(q_1 - q_{\alpha\beta})\delta(q_2 - q_{\alpha\gamma})\delta(q_3 - q_{\beta\gamma})} \tag{5.1}$$

In terms of the RSB scheme the calculation of this function is quite similar to that for the function $P(q)$. In particular, in terms of the replica matrix Q_{ab} instead of Eq. (4.16), in the present case one can easily prove that

$$P(q_1, q_2, q_3) = \lim_{n\to 0} \frac{1}{n(n-1)(n-2)} \sum_{a\neq b\neq c} \delta(Q_{ab} - q_1)\delta(Q_{ac} - q_2)\delta(Q_{bc} - q_3) \tag{5.2}$$

In terms of the Fourier transform of the function $P(q_1, q_2, q_3)$:

$$g(y_1, y_2, y_3) = \int dq_1 dq_2 dq_3 P(q_1, q_2, q_3) \exp\left[iq_1 y_1 + iq_2 y_2 + iq_3 y_3\right] \tag{5.3}$$

instead of Eq. (5.2) one gets:

$$g(y_1, y_2, y_3) = \lim_{n\to 0} \frac{1}{n(n-1)(n-2)} \sum_{a\neq b\neq c} \exp\left[iQ_{ab}y_1 + iQ_{ac}y_2 + iQ_{bc}y_3\right]$$
$$= \lim_{n\to 0} \frac{1}{n(n-1)(n-2)} \mathrm{Tr}\left[\hat{A}(y_1)\hat{A}(y_2)\hat{A}(y_3)\right] \tag{5.4}$$

where

$$A_{ab}(y) = \begin{cases} \exp(iQ_{ab}y); & a \neq b \\ 0; & a = b \end{cases} \tag{5.5}$$

Let us substitute the RSB solution for the matrix Q_{ab} into Eq. (5.4). In the continuum RSB limit the matrix Q_{ab} turns into the function $q(x)$, and according to the Parisi algebra (Section 3.4) the replica matrix $A_{ab}(y)$ turns into the corresponding function $A(x;y)$:

$$A(x;y) = \exp\{iq(x)y\} \tag{5.6}$$

Using the algorithms of the Parisi algebra, Eqs. (3.39)–(3.43), after simple calculations one obtains:

$$\lim_{n\to 0}\left(\frac{1}{n(n-1)(n-2)} \mathrm{Tr}\left[\hat{A}(y_1)\hat{A}(y_2)\hat{A}(y_3)\right]\right)$$
$$= \frac{1}{2}\int_0^1 dx \left[xA(x;y_1)A(x;y_2)A(x;y_3) + A(x;y_1)\int_0^x dz A(z;y_2)A(z;y_3)\right.$$
$$\left. + A(x;y_2)\int_0^x dz A(z;y_1)A(z;y_3) + A(x;y_3)\int_0^x dz A(z;y_1)A(z;y_2)\right] \tag{5.7}$$

Accordingly, for the function $P(q_1, q_2, q_3)$:

$$P(q_1, q_2, q_3) = \int dy_1 dy_2 dy_3 \, g(y_1, y_2, y_3) \exp[-iq_1y_1 - iq_2y_2 - iq_3y_3] \tag{5.8}$$

one gets:

$$P(q_1, q_2, q_3) = \frac{1}{2}\int_0^1 dx \left[x\delta(q(x) - q_1)\delta(q(x) - q_2)\delta(q(x) - q_3)\right.$$
$$+ \delta(q(x) - q_1)\int_0^x dz\delta(q(z) - q_2)\delta(q(z) - q_3)$$
$$+ \delta(q(x) - q_2)\int_0^x dz\delta(q(z) - q_1)\delta(q(z) - q_3)$$
$$\left. + \delta(q(x) - q_3)\int_0^x dz\delta(q(z) - q_1)\delta(q(z) - q_2)\right] \tag{5.9}$$

Introducing the integration over q instead of that over x and taking into

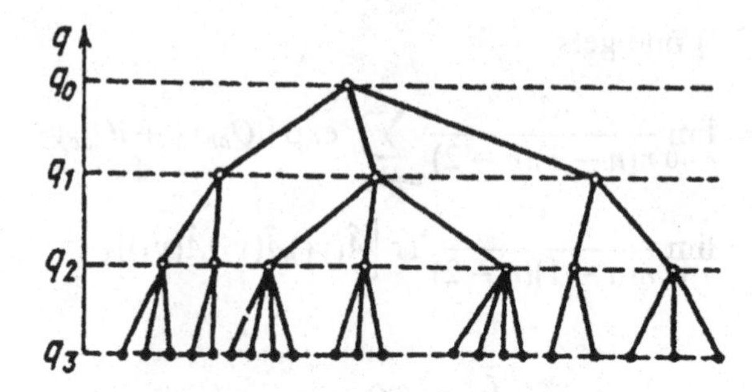

Fig. 5.1. The ultrametric tree of the spin-glass states.

account that $dx(q)/dq = P(q)$ one finally obtains the following result:

$$\begin{aligned} P(q_1, q_2, q_3) = & \frac{1}{2} P(q_1) x(q_1) \delta(q_1 - q_2) \delta(q_1 - q_3) \\ & + \frac{1}{2} P(q_1) P(q_2) \theta(q_1 - q_2) \delta(q_2 - q_3) \\ & + \frac{1}{2} P(q_2) P(q_3) \theta(q_2 - q_3) \delta(q_3 - q_1) \\ & + \frac{1}{2} P(q_3) P(q_1) \theta(q_3 - q_1) \delta(q_1 - q_2) \end{aligned} \tag{5.10}$$

From this equation one can easily see the following crucial property of the function $P(q_1, q_2, q_3)$. It is non-zero only in the following three cases: $q_1 = q_2 \leq q_3$; $q_1 = q_3 \leq q_2$; $q_3 = q_2 \leq q_1$. In all other cases the function $P(q_1, q_2, q_3)$ is identically equal to zero. In other words, this function is not equal to zero only if at least two of the three overlaps are equal, and their value is not bigger than the third one. It means that in the space of spin-glass states there exist no triangles with all three sides being different. The spaces having the above metric property are called *ultrametric*.

A simple illustration of the ultrametric space can be given in terms of the hierarchical tree (Fig. 5.1). The ultrametric space here is associated with the set of the endpoints of the tree. By definition, the overlaps between any two points of this space depend only on the number of 'generations' (in the 'vertical' direction) to the level of the tree where these two points have a common ancestor. One can easily check that paired overlaps among three arbitrary points of this set do satisfy the above ultrametric property.

The reader can find a detailed description of the ultrametric spaces in the review [3]. Here we are going to concentrate only on the general qualitative properties of the ultrametricity that are crucial for the physics of the spin-glass state.

5.2 Tree of states

Let us consider how the spin-glass ultrametric structures can be defined in more general terms.

Consider the following discrete stochastic process, which is assumed to take place *independently* at each site i of the lattice.

(1) At the first step, with the probability $P_0(y)$ one generates n_1 random numbers y^{α_1} ($\alpha_1 = 1, 2, \ldots, n_1$), which belong to the interval $[-1, +1]$.
(2) At the second step, for each y^{α_1} with the conditional probability $P_1(y^{\alpha_1}|y)$ one generates n_2 random numbers $y^{\alpha_1\alpha_2}$ ($\alpha_2 = 1, 2, \ldots, n_2$), belonging to the same interval $[-1, +1]$.
(3) At the third step, for each $y^{\alpha_1\alpha_2}$ with the conditional probability $P_2(y^{\alpha_1\alpha_2}|y)$ one generates n_3 random numbers $y^{\alpha_1\alpha_2\alpha_3}$ ($\alpha_3 = 1, 2, \ldots, n_3$), belonging to the same interval $[-1, +1]$, etc.

This process is continued up to the L-th step. Finally, in the interval $[-1, +1]$ one gets $n_1 n_2 \ldots n_L$ random numbers, which are described by the following set of the probability functions

$$P_{l-1}(y^{\alpha_1\ldots\alpha_{l-1}}|y^{\alpha_1\ldots\alpha_l}) \qquad (l = 1, 2, \ldots, L) \tag{5.11}$$

This stochastic (Markov) process takes place independently at each site of the lattice. Then, for each set of the obtained random numbers let us define the corresponding site spin states as follows:

$$\sigma_i^{\alpha_1\ldots\alpha_L} = \text{sign}\,(y_i^{\alpha_1\ldots\alpha_L}) \tag{5.12}$$

This way one obtains the set of $n_1 n_2 \ldots n_L$ spin states, which are labelled by the hierarchical 'address' $\alpha_1 \ldots \alpha_L$. The 'address' of a concrete state describes its genealogical 'history'.

Simple probabilistic arguments show that the overlap between any two spin states depends only on the degree of their 'relativeness', i.e. it is defined only by the number of generations that separate them from the closest common ancestor. Consider two spin states that have the following 'addresses':

$$\alpha_1\alpha_2 \ldots \alpha_l\alpha_{l+1}\alpha_{l+2} \ldots \alpha_L$$

and

$$\alpha_1\alpha_2 \ldots \alpha_l\beta_{l+1}\beta_{l+2} \ldots \beta_L$$

where $\beta_{l+1} \neq \alpha_{l+1}$. These two 'addresses' are becoming different, starting from the generation number l. As the stochastic processes generating the states are independent at each site, for the overlap between these two states

$$q^{\alpha_1...\alpha_l\alpha_{l+1}...\alpha_L}_{\alpha_1...\alpha_l\beta_{l+1}...\beta_L} = \frac{1}{N}\sum_i^N \sigma_i^{\alpha_1...\alpha_l\alpha_{l+1}...\alpha_L}\sigma_i^{\alpha_1...\alpha_l\beta_{l+1}...\beta_L} \tag{5.13}$$

in the thermodynamic limit $N \to \infty$ one gets:

$$q^{\alpha_1...\alpha_l\alpha_{l+1}...\alpha_L}_{\alpha_1...\alpha_l\beta_{l+1}...\beta_L} = \int_{-1}^{+1} dy_1 \dots dy_l P_0(y_1)P_1(y_1|y_2)\dots P_{l-1}(y_{l-1}|y_l)$$

$$\times \left[\int_{-1}^{+1} dy_{l+1}\dots dy_L P_l(y_l|y_{l+1})P_{l+1}(y_{l+1}|y_{l+2})\dots P_{L-1}(y_{L-1}|y_L)\,\mathrm{sign}\,(y_L)\right]^2 \equiv q_l \tag{5.14}$$

Therefore, the overlap depends only on the number l of the level of the tree at which the two states were separated in their genealogical history, and does not depend on the concrete 'addresses' of these states. One can easily see that it automatically means that the considered set of the states is ultrametric.

Note, that this is a general property of the considered stochastic evolution process, and it remains true for any choice of the probability distribution functions (5.11) which describe the concrete tree of states. A general reason for that is very simple. The above stochastic procedure has been defined as the random branching process that takes place in the infinite dimensional space (in the limit $N \to \infty$), and it is clear that here the branches, once separated, never come close again. Therefore, it is of no surprise that ultrametricity is observed in Nature very often. Examples are the space of biological species, the hierarchical state structures of disordered human societies, etc.

Let us consider the above hierarchical tree of states in some more detail. The equations for the overlaps between two spin states (5.13) and (5.14) can also be represented in terms of the so-called *ancestor states* $m^{\alpha_1...\alpha_l}$:

$$q_l = \frac{1}{N}\sum_i^N (m_i^{\alpha_1...\alpha_l})^2 \tag{5.15}$$

where the site magnetizations in the ancestor state $m^{\alpha_1...\alpha_l}$ at the level l are defined as follows:

$$m_i^{\alpha_1...\alpha_l} = \langle\sigma_i^{\alpha_1...\alpha_l\alpha_{l+1}...\alpha_L}\rangle_{(\alpha_{l+1}...\alpha_L)} \equiv m_l(y_i^{\alpha_1...\alpha_l}) \tag{5.16}$$

Here $\langle\dots\rangle_{(\alpha_{l+1}...\alpha_L)}$ denotes the averaging over all the descendant states (branches) of the tree outgoing from the branch $\alpha_1\dots\alpha_l$ at the level number

l. By definition:

$$m_l(y_i^{\alpha_1...\alpha_l}) = \int_{-1}^{+1} dy_{l+1}\dots dy_L P_l(y_i^{\alpha_1...\alpha_l}|y_{l+1}) \\ \times P_{l+1}(y_{l+1}|y_{l+2})\dots P_{L-1}(y_{L-1}|y_L)\,\mathrm{sign}\,(y_L) \qquad (5.17)$$

This equation for the function $m_l(y)$ can also be written in the following recurrent form:

$$m_l(y) = \int_{-1}^{+1} dy' \mathbf{P}_{ll'}(y|y')m_{l'}(y') \qquad (5.18)$$

where

$$\mathbf{P}_{ll'}(y|y') = \int_{-1}^{+1} dy_{l+1}\dots dy_{l'-1}P_l(y|y_{l+1})P_{l+1}(y_{l+1}|y_{l+2})\dots P_{l'-1}(y_{l'-1}|y') \qquad (5.19)$$

Therefore, all the concrete properties of the tree of states, and in particular the values of the overlaps $\{q_l\}$, are fully determined by the set of the probability functions (5.11) or (5.19). For the complete description of a concrete spin-glass system all these functions have to be calculated, or at least the algorithms of their calculations must be derived. In particular, this can be done for the SK model of a spin-glass. Unfortunately, the corresponding calculations for this model are rather cumbersome, and the reader interested in the details may refer to the original papers [23] and [24]. Here only the final results will be presented.

The ultrametric tree of states that describes the spin-glass phase of the SK model is defined by the random branching process described above, in which the continuous limit $L \to \infty$ must be taken. In this limit, instead of the integer numbers l, which define the discrete levels of the hierarchy, it is more convenient to describe the tree in terms of the selfoverlaps $\{q_l\}$ of the ancestor states. In the limit $L \to \infty$ the discrete parameters $\{q_l\}$ turn into the continuous variable $0 \leq q \leq 1$.

Instead of the discrete 'one-step' functions (5.11) in the continuous limit it is more natural to describe the tree in terms of the functions (5.19) which define the evolution of the tree from the level q to the other level q'. It can be proved (and it is this proof which requires us to go through somewhat painful algebra) that in the continuous limit these functions are defined by the following non-linear diffusion equation:

$$-\frac{\partial}{\partial q}\mathbf{P} = \frac{1}{2}\frac{\partial^2}{\partial y^2}\mathbf{P} + x(q)m_q(y)\frac{\partial}{\partial y}\mathbf{P} \qquad (5.20)$$

with the initial condition:

$$\lim_{q \to q'} \mathbf{P}_{qq'}(y|y') = \delta(y - y') \tag{5.21}$$

Here $x(q)$ is the function inverse to $q(x)$ (which is given by the RSB solution, Chapter 3), and the function $m_q(y)$ is the continuous limit of the discrete function (5.18). It can be shown that this function defines the distribution of the site magnetizations in the ancestor states at the level q of the tree. One can easily derive from Eqs. (5.18) and (5.20) that the function $m_q(y)$ satisfies the following equation:

$$-\frac{\partial}{\partial q} m_q(y) = \frac{1}{2} \frac{\partial^2}{\partial y^2} m_q(y) + x(q) m_q(y) \frac{\partial}{\partial y} m_q(y) \tag{5.22}$$

The above equations fully describe the properties of the ultrametric tree of the spin-glass states of the SK model.

5.3 Scaling in the space of spin-glass states

Let us summarize all the results obtained for the spin-glass model with the long-range interactions:

(1) In terms of the formal replica calculations, the free energy of the system can be represented in terms of the functional $F[\hat{Q}]$, which depends on the $n \times n$ replica matrix $\hat{Q}$, (Section 3.1). In the thermodynamic limit the leading contribution to the free energy comes from the matrices $\hat{Q}^*$, which correspond to the extrema of this functional, and the physical free energy is obtained in the limit $n \to 0$. In this limit the extrema matrices $\hat{Q}^*$ are defined by the infinite set of parameters that can be described in terms of the continuous Parisi function $q(x)$ defined at the interval $0 \leq x \leq 1$ (Sections 3.3–3.4). In the low-temperature region near the phase-transition point this function can be obtained explicitly (Section 3.5, Fig. 4.1).

(2) On the other hand, in terms of qualitative physical arguments, one can define as the order parameter the distribution function $P(q)$, which gives the probability of finding a pair of pure spin-glass states having the overlap equal to q. In terms of the RSB scheme one can show that the distribution function $P(q)$ is defined by the Parisi function $q(x)$: $P(q) = dx(q)/dq$, where $x(q)$ is the inverse function to $q(x)$ (Section 4.2). The low-temperature solutions for $q(x)$ and for $P(q)$ show that there exists the continuous spectrum of the overlaps among the pure states.

(3) Next, one can introduce the 'three-point' distribution function $P(q_1, q_2, q_3)$ which gives the probability that three arbitrary pure states have their mutual pair overlaps equal to q_1, q_2 and q_3. In terms of the RSB scheme this function can be calculated explicitly, and the obtained result shows that the space of the pure states has the ultrametric topology (Section 5.1).

(4) It can be shown that the ultrametric tree-like structures can be described in terms of the hierarchical evolution tree, which is defined by the random branching process.

Based on the above results, the spin-glass phase can be described in the qualitative physical terms as follows (see also Chapter 2).

At a given temperature T below T_c, the space of spin states is split into numerous pure states (valleys) separated by infinite energy barriers. Although the average site magnetizations m_i are different in different states, the value of the selfoverlap:

$$q(T) = \sum_i^N m_i^2 \tag{5.23}$$

appears to be the same in all the states. The value of q is the function of the temperature ($q(T_c) = 0$; $q(0) = 1$), and near T_c it can be calculated explicitly.

On the other hand, the overlaps $q^{\alpha\beta}$ of the pure states cover continuously the whole interval $0 \leq q^{\alpha\beta} \leq q(T)$. (In the presence of an external magnetic field h this interval starts from a non-zero value: $q_0(h,T) \leq q^{\alpha\beta} \leq q_1(h,T)$.) The distribution of the values of the overlaps $q^{\alpha\beta}$ is described by a probability function $P(q)$, which depends on the temperature (and the magnetic field). The structure of the space of the pure states can be described in terms of the ultrametric hierarchical tree discussed above.

Now, if the temperature is slightly decreased $T \to T' = T - \delta T$, each of the pure states is split into numerous new 'descendant' pure states. These states are characterized by the new value of the selfoverlap $q(T') > q(T)$. Correspondingly, the interval of their overlaps becomes bigger: $0 \leq q^{\alpha\beta} \leq q(T')$.

At a further decrease of the temperature, each of the newly born pure states is split again into new descendant pure states, and this branching process continues down to zero temperature ($q(T \to 0) \to 1$). The tree of pure states obtained this way has the property of the self-similarity (scaling), and at any given temperature the natural scale in the space of states is given by the value of $q(T)$.

Owing to infinite energy barriers separating the valleys, the 'observable' physics at the given temperature T is defined by only one of the pure states, which in terms of the hierarchical tree corresponds to one of the 'ancestor' states at the level (scale) $q(T)$. All these states are revealed in the horizontal cross-section of the tree at the level $q(T)$.

5.4 Phenomenological dynamics

Although the dynamical properties of spin glasses is an extremely hard problem even at the mean-field level, certain (the most simple) general slow relaxation properties of the disordered systems with the hierarchical structure of the free energy landscape can be understood rather easily using a purely phenomenological approach [25].

Assume that the free energy landscape in the spin-glass phase is of the type shown in Fig. 2.2: big wells contain a lot of smaller ones, each of the smaller wells contains a lot of even smaller ones, and so on. This kind of landscape could be characterized by the typical value of the *finite* energy barrier $\Delta(q)$ separating the wells at the scale q. Assuming that this landscape has the scaling property, the dependence of the typical value of the energy barrier Δ from the scale q could be described by the following simple scaling law:

$$\Delta(q) = \Delta_0(q - q(T))^{-\nu}; \qquad (q > q(T);\ \nu > 0) \tag{5.24}$$

Here $q(T)$ is the value of the selfoverlap of the pure states at the temperature T. The parameter $q(T)$ is the characteristic scale (the typical scale of the valleys) at which the barriers separating the states are becoming infinite.

Consider now what kind of relaxation properties could be derived from the above assumptions. The characteristic time needed to overcome the barrier Δ is

$$\tau(\Delta) \sim \tau_0 \exp\left(\frac{\Delta}{T}\right) \tag{5.25}$$

where τ_0 is characteristic microscopic time. Thus, the spectrum of the relaxation times inside the valley can be represented as follows:

$$\tau(q) \sim \tau_0 \exp[\beta\Delta_0(q - q(T))^{-\nu}] \tag{5.26}$$

Then the long-time relaxation behavior of the order parameter

$$q(t) = \frac{1}{N}\sum_i \langle\sigma_i(0)\sigma_i(t)\rangle \tag{5.27}$$

can be estimated (very roughly) as follows:

$$q(t) \sim \int_{q(T)}^{1} dq\, q \exp\left(-\frac{t}{\tau(q)}\right) \tag{5.28}$$

Using (5.26), one gets:

$$q(t) \sim \int_{q(T)}^{1} dq \exp\left(\ln(q) - \frac{t}{\tau_0} \exp[-\beta\Delta_0(q - q(T))^{-\nu}]\right) \tag{5.29}$$

In the limit of large times $t \gg \tau_0$ the saddle-point estimate of the above integral gives the following result:

$$q(t) \sim q(T) + \left[\frac{\beta\Delta_0}{\ln(t/\tau_0)}\right]^{1/\nu} \tag{5.30}$$

Therefore at large times the order parameter approaches its equilibrium value $q(T)$ logarithmically slowly. Apparently, the relaxation behavior of other observable quantities should be of the same slow type.

Of course, true dynamic properties of spin glasses are much more complicated, and they can not be reduced only to the phenomenon of extremely slow relaxation. Actually, the main property of spin glasses is that they can not reach true thermodynamic equilibrium at any finite observation time (the reader can find detailed discussion of this problem in a recent review paper [8]). Because the theoretical achievements in understanding the dynamical properties of spin glasses are far from being impressive yet, in the next chapter we will consider the results of experimental observations of the relaxation phenomena in real spin-glass magnets.

6

Experiments

In this chapter we will consider classical experiments that have been performed on *real* spin glass materials, aiming to check to what extent the qualitative picture of the spin-glass state described in previous chapters does take place in the real world. The main problem of the experimental observations is that the concepts and quantities that are very convenient in theoretical considerations are rather far from the experimental realities, and it is a matter of the experimental art to invent convincing experimental procedures that would be able to confirm (or reject) the theoretical predictions.

A series of such brilliant experiments has been performed by M. Ocio, J. Hammann, F. Lefloch and E. Vincent (Saclay), and M. Lederman and R. Orbach (UCLA) [9]. Most of these experiments have been done on the crystals $CdCr_{1.7}In_{0.3}S_4$. The magnetic disorder there is present due to the competition of the ferromagnetic nearest neighbor interactions and the antiferromagnetic higher-order neighbor interactions. This magnet has already been systematically studied some time ago [26], and its spin-glass phase transition point $T = 16.7\,\mathrm{K}$ is well established. Some of the measurements have been also performed on the metallic spin glasses AgMn [27] and the results obtained were qualitatively quite similar. It indicates that presumably the qualitative physical phenomena observed do not depend very much on the concrete realization of the spin-glass system.

6.1 Aging

The phenomenon of *aging* in spin glasses has been known for many years [28]. It is not directly connected with the hierarchy of the spin-glass states, but it explicitly demonstrates the absence of true thermodynamic equilibrium in spin glasses.

The procedure of the experiments is as follows. The sample is cooled

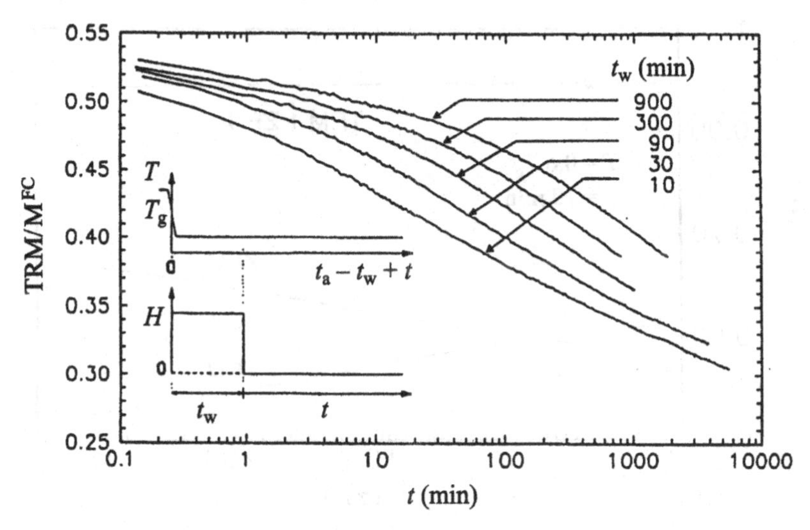

Fig. 6.1. The relaxation behavior of the magnetization in the field cooled aging experiments.

down into the spin-glass state in the presence of a weak uniform magnetic field H. Then, at a constant temperature $T < T_c$ the sample is kept in this magnetic field during some waiting time t_w. Finally the magnetic field is switched off, and the measurement of the relaxation of the thermoremainent magnetization (TRM) is performed. The results of these measurements for different values of t_w are shown in Fig. 6.1.

The first important result of these measurements is that the observed relaxation is extremely slow and non-exponential (note, that the typical values of t_w are macroscopic: minutes, hours, days). More important, however, is that the relaxation appears to be *non-stationary*: the relaxation processes that take place in the system *after* switching off the field depend on the 'lifetime' t_w of the system *before* the measurement was started. The spin glass is getting 'stiffer' with the time: the bigger t_w is, the slower the relaxation goes on. Therefore, the results of the measurements depend on two time scales: the observation time t, and the time that has passed after the system came into the spin-glass state, the 'aging' time t_w. It is crucial that at all experimentally accessible time scales no indication that the relaxation curves are reaching saturation at some limiting curve (corresponding to $t_w = \infty$) has been observed. Thus, at any experimentally accessible times such a system remains out of the true thermal equilibrium.

Note that it is not the presence of the magnetic field that is responsible for the observed phenomenon. The magnetic field here is just the instrument that makes it possible to demonstrate it. One can also perform the 'mirror'

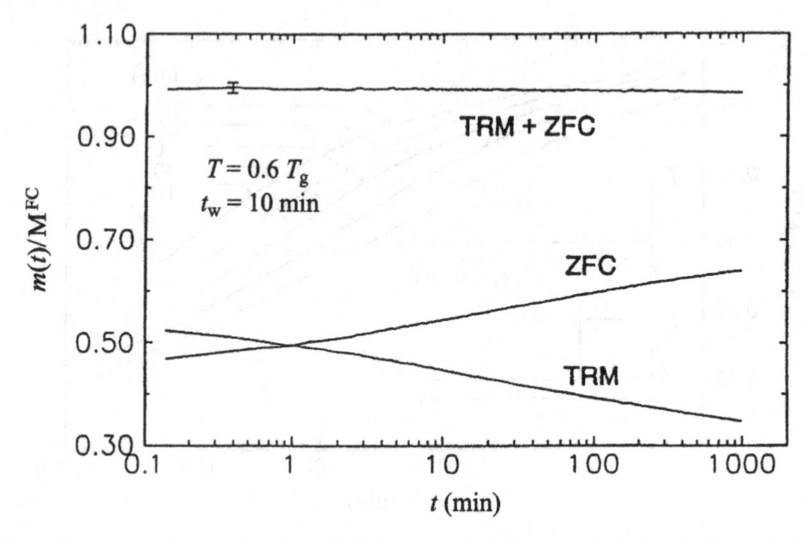

Fig. 6.2. The relaxation behavior of the magnetization in the zero field cooled aging experiments.

experiment: the system is cooled down into the spin-glass state in the zero magnetic field, then it is kept at a constant temperature $T < T_c$ during some waiting time t_w, and finally the magnetic field is switched on and the relaxation of the magnetization is measured. Again, the results of the measurements essentially depend on t_w. Moreover, for any given value of t_w the curves obtained in these two types of the experiment appear to be symmetric: the sum of the values of the magnetizations obtained in these 'mirror' experiments appears to be a time-independent constant (Fig. 6.2).

6.2 Temperature cycles and the hierarchy of states

Now we consider two types of the experiments which are supposed to reveal the effects connected with the existence of the hierarchical tree of spin-glass states.

In the experiments of the first type, the sample, in a weak magnetic field, is cooled down into the spin-glass phase. Then, it is kept at a constant temperature $T < T_c$ during some waiting time t_{w_1}. After that the temperature is slightly changed down to $T' = (T - \Delta T)$ (where the value of ΔT is small), and the sample is kept at this temperature during waiting time t_{w_3}. Then the temperature is changed up to the original value T again, and the sample is kept at this constant temperature during waiting time t_{w_2}. After that the magnetic field is switched off and the relaxation of the magnetization is measured. The results for different values of ΔT are shown in Fig. 6.3.

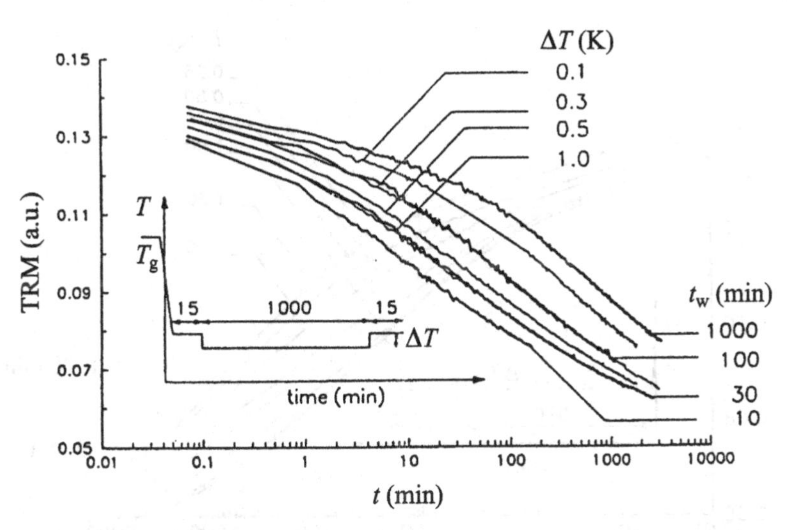

Fig. 6.3. The relaxation behavior of the magnetization in the aging experiments with the cooling temperature cycles.

The main result of these measurements is the following. It is clear from the plots of Fig. 6.3 that if the value of the temperature step ΔT is not too small, then all the relaxation curves appear to be *identical* to those in the usual aging experiments (Section 6.1) with the waiting time $t_{\mathrm{w}} = t_{\mathrm{w}_1} + t_{\mathrm{w}_2}$. It means that for the processes of equilibration at the temperature T, the system remained effectively completely frozen during all the time period t_{w_3} when it was kept at the temperature $(T - \Delta T)$.

In the experiments of the second type, again, the sample in the presence of a weak magnetic field is cooled down into the spin-glass phase, and then it is kept at a constant temperature $T < T_c$ during waiting time t_{w_1}. Next, the sample is slightly heated up to the temperature $T' = (T + \Delta T)$, (where the value of ΔT is small) and after a relatively short time interval it is cooled down again to the original temperature T. Then, it is kept at this constant temperature during waiting time t_{w_2}, and after that, the magnetic field is switched off and the relaxation of the magnetization is measured. The results for different values of ΔT are shown in Fig. 6.4.

In this case one finds that if the value of the temperature step ΔT is not too small, then all the relaxation curves appear to be *identical* to those in the usual aging experiments (Section 6.1) with the waiting time $t_{\mathrm{w}} = t_{\mathrm{w}_2}$. It means that even slight heating is enough to wipe out all the aging that has been 'achieved' at the temperature T during the time period before heating. In other words, after the slight heating jump the equilibration processes start all over again, while all the 'pre-history' of the sample appears to be wiped out. (Note that the temperature $(T + \Delta T)$ is still well below T_c.)

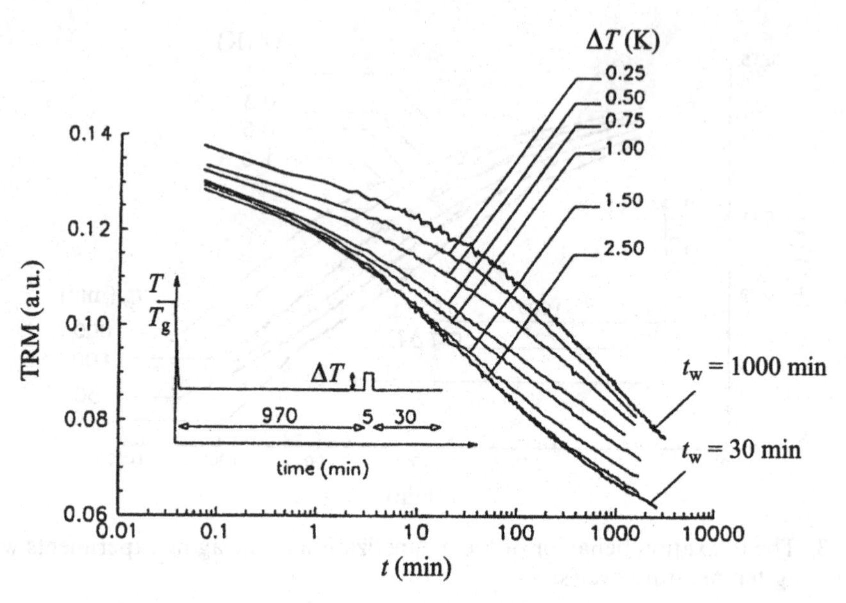

Fig. 6.4. The relaxation behavior of the magnetization in the aging experiments with the heating temperature cycles.

Such quite asymmetric response of the system with respect to the considered temperature cycles of cooling and heating can be well explained in terms of the qualitative physical picture of the continuous hierarchy of the phase transitions and the tree-like structure of the spin-glass states.

The qualitative interpretation of the results described above is as follows. The process of thermal equilibration, as time goes on, can be imagined as the process of jumping over higher and higher energy barriers in the space of states. After some waiting time t_w, the system covers a certain part of the configurational space, which could be characterized by the maximum energy barriers of the order of $\Delta_{max} \simeq T \log(t_w/\tau)$ (here τ is characteristic microscopic time). It is assumed that any scale in the configurational space is characterized by certain typical values of the energy barriers (see also Section 5.4). Then the results of the experiments with the temperature cycles of cooling can be interpreted as follows. During the time period t_{w_1} when the system is kept at the temperature T, it covers a certain finite part of the configurational space inside one of the valleys. After cooling down to the temperature $(T - \Delta T)$ this part of the configurational space is split into several smaller valleys separated by *infinite* energy barriers. Besides, the finite energy barriers separating the metastable states inside the valleys are getting higher, while some of these metastable states are split into many new ones. Then, during the time t_{w_3}, the system is trying to cover these new states being locked by infinite barriers in a limited part of the configurational space. Therefore, whatever time has passed at the temperature $(T-\Delta T)$ the system can cover only those states, which are the descendants of the states already occupied at the temperature T, and not more. Note that this phenomenon of *ergodicity breaking* is just the consequence of the *phase transition* that occurred in the system due to cooling down from the temperature T to the temperature $(T-\Delta T)$. Then, after heating back to the original temperature T all these descendant states are merging together into their ancestors, and the process of thermal equilibration at the temperature T continues again, as if there was no time interval that the system spent at the temperature $(T-\Delta T)$.

In the experiments with the temperature cycles of heating, the effects to be expected are different. After heating to the temperature $(T+\Delta T)$ the states occupied by the system during the time t_{w_1} at the temperature T merge together into the smaller number of their ancestor states. If the value of ΔT is not too small, such that $q(T+\Delta T) < q'$, where $q(T)$ is the selfoverlap of the states at the temperature T, and q' is the selfoverlap of the common ancestor of the states occupied during time interval t_{w_1}, then after heating all the occupied states would merge together into one common ancestor state. Within this limited part of the phase space this effectively corresponds to the paramagnetic phase transition. Therefore, all the thermal equilibration 'achieved' at the temperature T will be wiped out, and after cooling back to the original temperature T the process of thermal equilibration will start all over again.

In brief, the results of the experiments considered here could be summarized as follows. If the spin-glass system is equilibrating at some temperature $T < T_c$, then any temporary heating would eliminate all the equilibration achieved, while any cooling for any time period just postpones the equilibration processes at this temperature.

6.3 Temperature dependence of the energy barriers

The scheme of the above experiments can be slightly changed so that it would be possible to estimate the temperature dependence of the (finite) free energy barriers.

The experiments have been done on the metallic spin glasses AgMn $(T_c = 10.4\,\mathrm{K})$. The scheme of the experiments is as follows. First, the spin glass is aging in a weak magnetic field during the waiting time t_w at the temperature $(T-\Delta T)$. Then the sample is quickly heated up to the

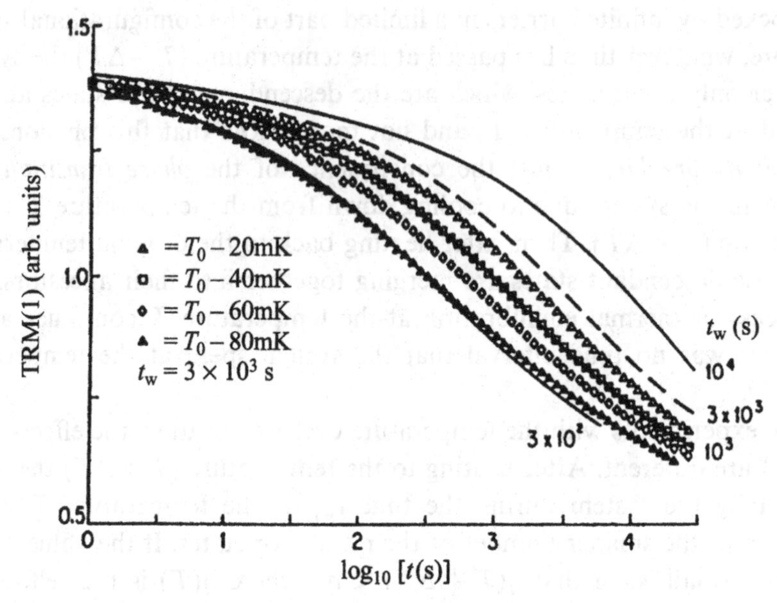

Fig. 6.5. The relaxation behavior of the magnetization at the temperature T after aging at the temperature $T - \Delta T$.

temperature T, and simultaneously the magnetic field is switched off. After that, the measurements of the relaxation of the magnetization are observed.

The results are shown in Fig. 6.5. These plots clearly show that if the value of ΔT is not too small, then the relaxation curves obtained are practically identical to those in the usual aging experiments at the same temperature T but with some other waiting time $t_w^{\text{eff}} < t_w$.

Assuming that the values of the *finite* energy barriers separating metastable states essentially depend on the temperature, this phenomenon can also be easily explained in terms of the hierarchical structure of the spin-glass states. Because the free energy barriers at the temperature $(T - \Delta T)$ must be higher than corresponding barriers at the temperature T, the region of the phase space occupied by the system at the temperature $(T - \Delta T)$ is bounded by barriers that are getting smaller at the temperature T. Correspondingly, the time needed to cover this part of the phase space at the temperature T is smaller than that at the temperature $(T - \Delta T)$. The crucial point is the following. At the initial moment of the measurements the values of the temperature and the magnetic field in these two types of the experiment are the same, and if the value of t_w^{eff} is chosen correctly, then the long-time relaxation curves obtained appear to be identical. It means that the region of the phase space occupied by the system at the initial moment of the measurements in both cases must be the same. If the system is equilibrating at the temperature T this region can be characterized by the maximum value of the typical energy barriers:

$$\Delta(T; t_W^{\mathrm{eff}}) = T \log\left(\frac{t_W^{\mathrm{eff}}}{\tau}\right) \tag{6.1}$$

Correspondingly, if the equilibration takes place at the temperature $(T - \Delta T)$, the typical value of the maximum barriers is:

$$\Delta(T - \Delta T; t_W) = (T - \Delta T) \log\left(\frac{t_W}{\tau}\right) \tag{6.2}$$

Because the relaxation processes both after the aging at temperature T during the time t_W and after the aging at the temperature $(T - \Delta T)$ during the time t_W^{eff} are the same, the initial state of the system must also be the same. Therefore, one can conclude that $\Delta(T - \Delta T)$ and $\Delta(T)$ are the heights of *the same* barrier at different temperatures. Based on this conclusion and using the experimental data of Fig. 6.5, one can get the plot for the dependence of the value $\partial\Delta/\partial T$ on Δ at the given temperature. In Fig. 6.6 the dependence of $\Delta(T - \Delta T)$ from $\Delta(T)$ is shown for $T = 9$ K, 9.5 K and 10 K at a fixed value of the temperature jump $\Delta T = 20$ mK. These plots demonstrate that within the experimental errors the dependencies obtained at different T coincide.

In Fig. 6.7 the corresponding dependence of the value $\partial\Delta/\partial T$ from Δ is shown. Within the experimental errors the value of $\partial\Delta/\partial T$ depends only on Δ and it does not depend directly on the temperature. The dashed line in Fig. 6.7 is the power law fitting of the experimental data:

$$\frac{d\Delta}{dT} \simeq a\Delta^6; \qquad a = 2.9 \times 10^{-7} \tag{6.3}$$

Integrating this equation, one gets:

$$\Delta(T) \simeq \left[\frac{T - T^*}{T_c}\right]^{-1/5}; \quad T > T^* \tag{6.4}$$

The temperature T^* is the integration constant, which actually labels the concrete barrier. In other words, each barrier can be characterized by the critical temperature T^* at which this (finite at $T > T^*$) barrier becomes infinite. In this sense the critical temperature T_c can be interpreted just as the maximum possible value of T^*.

In conclusion, the experiments considered above clearly demonstrate the absence of the thermal equilibrium in the spin-glass phase at all experimentally accessible time scales. These experiments also demonstrate the existence of the whole spectrum of the free energy barriers up to infinite values, at

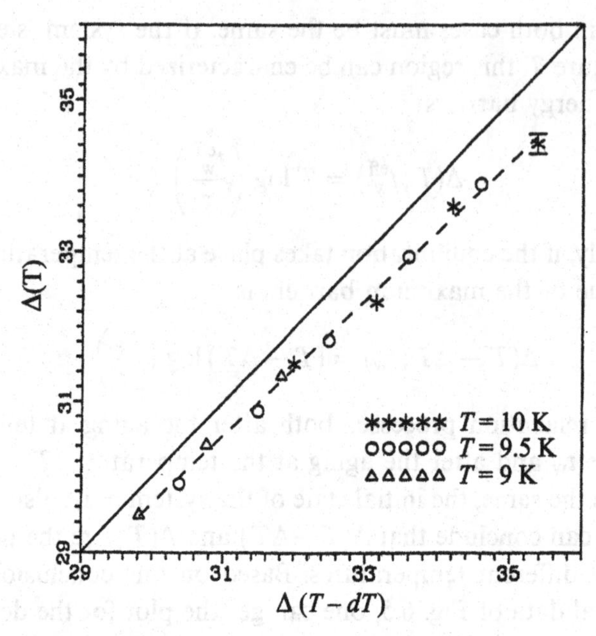

Fig. 6.6. The dependence of the values of the free energy barriers at the temperature T from their values at the temperature $T - \Delta T$.

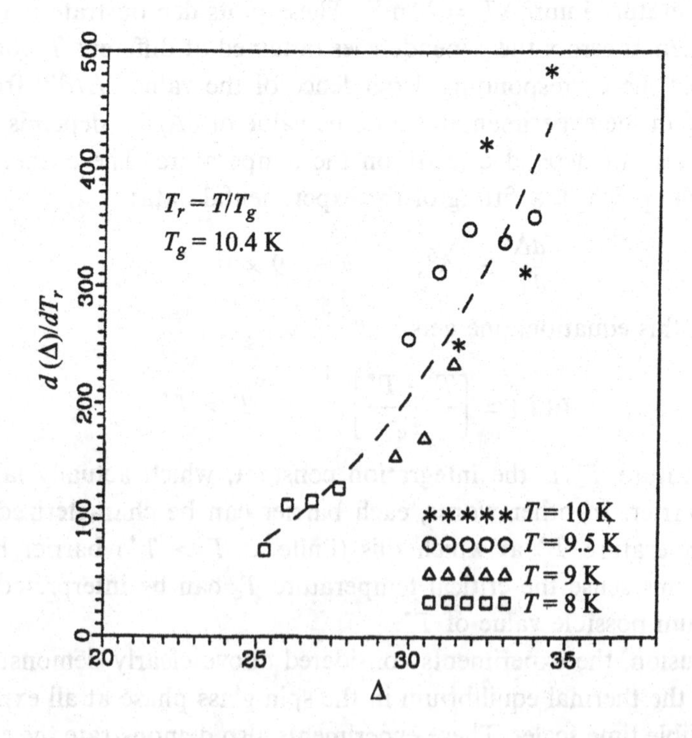

Fig. 6.7. The dependence of $d\Delta/dT$ from the values of the barriers Δ.

any temperature below T_c. The results of the measurements show that the barriers' heights strongly depend on the temperature, and at *any* temperature $T < T_c$ certain barriers are becoming infinite. This phenomenon clearly indicates the presence of the ergodicity breaking phase transition at any temperature below T_c, which results in the continuous process of fragmentation of the phase space into smaller and smaller valleys with decrease of the temperature.

Part two

Critical phenomena and quenched disorder

7

Scaling theory of the critical phenomena

7.1 The Ginzburg–Landau theory

We begin our study of the critical phenomena at phase transitions of the second order with the mean-field approximation discussed in the Introduction (Section 1.2). The starting point for further consideration is the mean-field expansion of the free energy in the vicinity of the critical point T_c, (Eq. (1.28), Fig. 1.1):

$$f(\phi) = \frac{1}{2}\tau\phi^2 + \frac{1}{4}g\phi^4 - h\phi \tag{7.1}$$

where $\tau = (T - T_c)/T_c \ll 1$ is the reduced temperature, h is the external magnetic field. Here the 'coupling constant' g is the parameter of the theory, and the order parameter $\phi = \langle\sigma_i\rangle$ is the average spin magnetization. The value of ϕ is determined from the condition of the minimum of the free energy, $df/d\phi = 0$:

$$\tau\phi + g\phi^3 = h \tag{7.2}$$

and $d^2f/d\phi^2 > 0$.

In the absence of the external magnetic field ($h = 0$) at temperatures above T_c, ($\tau > 0$), the free energy has only one (trivial) minimum at $\phi = 0$. Below the critical point, $\tau < 0$, the free energy has two minima, and the corresponding solutions of the saddle-point equation (7.2) are:

$$\phi(\tau) = \pm\sqrt{\frac{|\tau|}{g}} \tag{7.3}$$

As $T \to T_c$ from below, $\phi(T) \to 0$.

As has already been discussed in the Introduction, this very simple mean-field theory demonstrates on a qualitative level the fundamental phenomenon called the spontaneous symmetry breaking. At the critical temperature $T =$

T_c the phase transition of the second order occurs, such that in the low-temperature region $T < T_c$ the symmetry with respect to the global change of the signs of the spins is broken, and *two* (instead of one) ground states appear. These two states differ by the sign of the average spin magnetization, and they are separated by the macroscopic barrier of the free energy.

In a small non-zero magnetic field ($h \ll 1$) the qualitative shape of the free energy is shown in Fig. 1.1(b). In this case the saddle-point equation (7.2) always has a non-zero solution for the order parameter ϕ at all temperatures. In particular, in the low-temperature region ($\tau < 0$) we find:

$$\phi \simeq \begin{cases} \sqrt{\frac{|\tau|}{g}} + \frac{h}{2\tau} & \text{for } h \ll h_c(\tau) \\ \left(\frac{h}{g}\right)^{1/3} & \text{for } h \gg h_c(\tau) \end{cases} \tag{7.4}$$

where

$$h_c(\tau) = \frac{1}{\sqrt{g}}|\tau|^{3/2} \tag{7.5}$$

whereas in the high-temperature region ($\tau > 0$):

$$\phi \simeq \begin{cases} \frac{h}{\tau} & \text{for } h \ll h_c(\tau) \\ \left(\frac{h}{g}\right)^{1/3} & \text{for } h \gg h_c(\tau) \end{cases} \tag{7.6}$$

Thus, at $h \neq 0$ the phase transition is 'smoothed out' in the temperature interval $|\tau| \sim h^{2/3}$ (Eq. (7.5)) around T_c.

The physical quantity, which describes the reaction of the system on the infinitely small magnetic field, is called susceptibility. It is defined as follows:

$$\chi = \frac{\partial \phi}{\partial h}|_{h=0} \tag{7.7}$$

According to Eqs. (7.4)–(7.6) one finds that near the critical point the susceptibility becomes divergent:

$$\chi \simeq \begin{cases} \tau^{-1} & \text{at } T > T_c \\ \frac{1}{2}|\tau|^{-1} & \text{at } T < T_c \end{cases} \tag{7.8}$$

For the so called non-linear susceptibility $\chi(h) = \partial\phi/\partial h$ in the close vicinity of the critical point ($|\tau|^{3/2} \ll h\sqrt{g}$), we get:

$$\chi(h) \simeq h^{-2/3} \tag{7.9}$$

The other basic physical quantity is the specific heat, which is defined as follows:

$$C = -T\frac{\partial^2 f}{\partial T^2} \tag{7.10}$$

For the specific heat near the critical point (in the zero magnetic field), according to Eqs. (7.3)–(7.1) we obtain:

$$C \simeq \begin{cases} 0 & \text{at } T > T_c \\ \text{const} = \dfrac{1}{2g} & \text{at } T < T_c \end{cases} \tag{7.11}$$

Of course, all the above results, which were obtained in terms of a very primitive mean-field approximation, cannot pretend to be reliable. Nevertheless, on a qualitative level they demonstrate a very important physical phenomenon: near the point of the second-order phase transition at least some of the physical quantities become singular (or non-analytic). Now let us consider one simple and natural improvement of the mean-field theory considered above.

An apparent defect of the mean-field approximation given above is that it does not take into account correlations among spins. This could be easily amended if we are interested in the studies of only *large-scale* phenomena, which will be shown to be responsible for the leading singularities in the thermodynamical functions. In this case the order parameters ϕ_i are almost spatially homogeneous, and they can be represented as slowly varying (with small gradients) functions of the continuous space coordinates. Then, the interaction term in the exact lattice Hamiltonian (1.14) can be approximated as follows:

$$\frac{1}{2}\sum_{<i,j>} \phi_i\phi_j \to \frac{1}{2}\int d^D x \left[\phi^2(x) + (\nabla\phi(x))^2\right] \tag{7.12}$$

Correspondingly, the Hamiltonian in which only small spatial fluctuations of the order parameter are taken into account can be written as follows:

$$H = \int d^D x \left[\frac{1}{2}(\nabla\phi(x))^2 + \frac{1}{2}\tau\phi^2(x) + \frac{1}{4}g\phi^4(x) - h\phi(x)\right] \tag{7.13}$$

The theory that is based on the above Hamiltonian is called the Ginzburg–Landau approach. In fact the Ginzburg–Landau Hamiltonian is nothing but the first few terms of the expansion in powers of ϕ and $(\nabla\phi)$. In the vicinity of the (second-order) phase-transition point, where the order parameter is small and the leading contributions come from large-scale fluctuations, such an approach looks quite natural.

Consider the contributions caused by small fluctuations at the background of the homogeneous order parameter $\phi_0 = \sqrt{|\tau|/g}$:

$$\phi(x) = \phi_0 + \varphi(x) \tag{7.14}$$

where $\varphi(x) \ll \phi_0$.

For simplicity let us consider the case of the zero magnetic field. Then the expansion of the Hamiltonian (7.13) to the second order in φ yields:

$$H = H_0 + \int d^D x \left[\frac{1}{2}(\nabla\varphi(x))^2 + |\tau|\varphi^2(x)\right] \tag{7.15}$$

In terms of the Fourier representation

$$\varphi(x) = \int \frac{d^D k}{(2\pi)^D}\varphi(k)\exp(-ikx) \tag{7.16}$$

one gets:

$$H = \frac{1}{2}\int \frac{d^D k}{(2\pi)^D}\left(k^2 + 2|\tau|\right) \mid \varphi(k) \mid^2 + H_0 \tag{7.17}$$

Therefore, for the correlation function

$$G_0(k) \equiv \langle|\varphi(k)|^2\rangle = \frac{\int D\varphi(k)|\varphi(k)|^2 \exp\left(-H[\varphi]\right)}{\int D\varphi(k)\exp\left(-H[\varphi]\right)} \tag{7.18}$$

one obtains the following result:

$$G_0(k) = \frac{1}{k^2 + 2|\tau|} \tag{7.19}$$

Besides, it is obvious that

$$\langle\varphi(k)\varphi(k')\rangle = G_0(k)\delta(k + k') \tag{7.20}$$

Therefore, for the spatial correlation function

$$\begin{aligned} G_0(x) &= \langle\langle\phi(0)\phi(x)\rangle\rangle \equiv \langle\phi(0)\phi(x)\rangle - \langle\phi(0)\rangle\langle\phi(x)\rangle \\ &= \langle\varphi(0)\varphi(x)\rangle = \int \frac{d^D k}{(2\pi)^D}\langle|\varphi(k)|^2\rangle \exp(ikx) \end{aligned} \tag{7.21}$$

we obtain:

$$G_0(x) \sim \begin{cases} |x|^{-(D-2)} & \text{for } |x| \ll R_c(\tau) = \dfrac{1}{\sqrt{2|\tau|}}; \quad (7.22a) \\ \exp\left(-|x|/R_c\right) & \text{for } |x| \gg R_c(\tau); \quad (7.22b) \end{cases}$$

Here the quantity

$$R_c(\tau) \sim |\tau|^{-\frac{1}{2}} \tag{7.23}$$

is called the correlation length.

Thus, the situation near T_c ($|\tau| \ll 1$) looks as follows. At scales much larger than the correlation length $R_c(\tau) \gg 1$ the fluctuations of the field $\phi(x)$ around its equilibrium value ϕ_0 ($\phi_0 = 0$ at $T > T_c$, and $\phi_0 = \sqrt{|\tau|/g}$ at $T < T_c$) become effectively independent (their correlations decay exponentially, Eq. (7.22b). On the other hand, at scales much smaller than $R_c(\tau)$, in the so called fluctuation region, the fluctuations of the order parameter are strongly correlated, and their correlation functions exibit weak power-law decay, Eq. (7.22a). Therefore, inside the fluctuation region the gradient, or the fluctuation term of the Hamiltonian (7.13) becomes crucial for the theory. At the critical point the fluctuation region becomes infinite.

Let us estimate to what extent the above simple considerations are correct. The expansion (7.15) could be used and the result (7.22) is justified only if the characteristic value of the fluctuations φ are small in comparison with the equilibrium value of the order parameter ϕ_0. Because the correlation length R_c is the only relevant spatial scale that exists in the system near the phase transition point, the characteristic value of the fluctuations of the order parameter could be estimated as follows:

$$\overline{\varphi^2} \equiv \frac{1}{R_c^D} \int_{|x|<R_c} d^D x \langle \varphi(0)\varphi(x) \rangle \sim R_c^{-(D-2)} \tag{7.24}$$

The above simple mean-field estimates for the critical behavior are valid only if the value of $\overline{\varphi^2}$ is much smaller than the corresponding value of the equilibrium order parameter ϕ_0^2:

$$R_c^{-D+2} \ll \frac{|\tau|}{g} \tag{7.25}$$

Using (7.23) we find that this condition is satisfied if:

$$g|\tau|^{\frac{D-4}{2}} \ll 1 \tag{7.26}$$

Therefore, if the dimension of the system is bigger than 4, near the phase-transition point, $\tau \to 0$, the condition (7.26) is always satisfied. On the other hand, at dimensions $D < 4$ this condition is always violated near the critical point.

Thus, these simple estimates reveal the following quite important points.

(1) If the dimension of the considered system is bigger than 4, then its critical behavior in the vicinity of the second-order phase transition is successfully described by the mean-field theory.
(2) If the dimension of the system is less than 4, then, according to Eq. (7.26), the mean-field approach gives correct results only in the

range of temperatures not too close to T_c:

$$\tau \gg \tau_*(D, g) \equiv g^{\frac{2}{4-D}}, \qquad (\tau \ll 1) \tag{7.27}$$

(here it is assumed that $g \ll 1$, otherwise there would be no mean-field critical region at all). In the close vicinity of T_c, $|\tau| \ll \tau_*$, the other (non-Gaussian) type of the critical behavior can be expected to occur.

7.2 Critical exponents

In general, it is believed that critical behavior of the physical quantities near the phase-transition point can be described in terms of the so-called *critical exponents*. In particular, for the quantities considered above, the critical exponents are defined as follows:

$$
\begin{array}{llll}
\text{Correlation length:} & R_c \sim |\tau|^{-\nu} & \text{at } h \ll h_c(\tau) & \\
 & R_c \sim h^{-\mu} & \text{at } h \gg h_c(\tau) & \\
\text{Order parameter:} & \phi_0 \sim |\tau|^{\beta} & \text{at } h \ll h_c(\tau);\ \tau < 0 & \\
 & \phi_0 \sim h^{1/\delta} & \text{at } h \gg h_c(\tau) & \\
\text{Specific heat:} & C \sim |\tau|^{-\alpha} & \text{at } h \ll h_c(\tau) & (7.28) \\
\text{Susceptibility:} & \chi \sim |\tau|^{-\gamma} & \text{at } h \ll h_c(\tau) & \\
 & \chi \sim h^{1/\delta - 1} & \text{at } h \gg h_c(\tau) & \\
\text{Correlation function:} & G(x) \sim |x|^{-D+2-\eta} & \text{at } |x| \ll R_c &
\end{array}
$$

where the value of the critical field is $h_c(\tau) \sim |\tau|^{\nu/\mu}$ (this estimate follows from the comparison of the correlation lengths in small and in large fields). In fact, not all the critical exponents listed in Eq. (7.28) are independent. One can easily derive (see below) the following relations among them:

$$\alpha = 2 - D\nu \tag{7.29}$$

$$\delta = \frac{D + 2 - \eta}{D - 2 + \eta} \tag{7.30}$$

$$\gamma = (2 - \eta)\nu \tag{7.31}$$

$$2\beta = 2 - \gamma - \alpha \tag{7.32}$$

$$\mu = \frac{2}{D+2-\eta} \tag{7.33}$$

For seven exponents there exist five equations, which means that only two exponents are independent. In other words, to find all the critical exponents one needs to calculate only two of them.

In particular, the Ginzburg–Landau mean-field theory considered above gives: $\nu = 1/2$ and $\eta = 0$ (see Eqs. (7.22)–(7.23)). Using Eqs. (7.29)–(7.33) one easily finds the rest of the exponents: $\alpha = -(D-4)/2$, $\delta = (D+2)/(D-2)$, $\gamma = 1$, $\beta = (D-2)/4$, $\mu = 1/3$. These critical exponents fully describe the critical behavior of any scalar field D-dimensional system at $D \geq 4$.

Let us now derive the relations (7.29)–(7.33). According to the definition of specific heat:

$$C = -T\frac{\partial^2 f}{\partial T^2} \tag{7.34}$$

one gets:

$$C = \frac{1}{V}\int d^D x \int d^D x' \left[\langle\phi^2(x)\phi^2(x')\rangle - \langle\phi^2(x)\rangle\langle\phi^2(x')\rangle\right] \sim \frac{1}{R_c^D}\langle\Phi\rangle^2 \tag{7.35}$$

where

$$\Phi = \int_{|x|<R_c} d^D x \phi^2(x) \tag{7.36}$$

According to Eq. (7.13), the equilibrium energy density of the system (at scales bigger than R_c) is proportional to $|\tau|\Phi$. Thus, the equilibrium value of $\langle\Phi\rangle$ is defined by the condition $|\tau|\langle\Phi\rangle \sim T$ ($T \simeq T_c = 1$ in our case). Therefore, from Eq. (7.35) one gets:

$$C \sim R_c^{-D}|\tau|^{-2} \sim |\tau|^{D\nu-2} \tag{7.37}$$

Thus, according to the definition of the critical exponent α, $C \sim |\tau|^{-\alpha}$, one obtains Eq. (7.29).

Using the definitions of the susceptibility, as well as the critical exponents of the correlation function η and that of the correlation length ν (Eq. (7.28)), one obtains:

$$\chi = \frac{\partial\langle\phi\rangle}{\partial h}|_{h=0} = \int d^D x \langle\langle\phi(0)\phi(x)\rangle\rangle \sim R_c^D\, R_c^{2-D-\eta} \sim |\tau|^{-\nu(2-\eta)} \tag{7.38}$$

On the other hand: $\chi \sim |\tau|^{-\gamma}$, which provides Eq. (7.31).

The value of the susceptibility, Eq. (7.38), can be estimated in another way:

$$\chi \sim R_c^D \phi_0^2 \sim |\tau|^{-D\nu+2\beta} \tag{7.39}$$

This yields: $\gamma = D\nu - 2\beta$. Using Eq. (7.29), one gets Eq. (7.32).

Now let us define the value of the order parameter in the region, which is less than the correlation length:

$$\psi \equiv \int_{|x|<R_c} d^D x \phi(x) \tag{7.40}$$

The characteristic value of the field ψ is:

$$\psi_c \equiv \sqrt{\langle \psi^2 \rangle} \sim \left(R_c^D \int_{|x|<R_c} d^D x \langle \phi(0)\phi(x) \rangle \right)^{1/2} \sim R_c^{\frac{D+2-\eta}{2}} \tag{7.41}$$

The critical value of the external field $h_c(\tau)$ is defined by the condition:

$$\psi_c h_c \sim T = 1 \tag{7.42}$$

Therefore, at this value of the field:

$$R_c(h) \sim h^{-\frac{2}{D+2-\eta}} \tag{7.43}$$

which yields Eq. (7.33).

On the other hand: $\psi_c \sim \phi_0 R_c^D$. Using the condition (7.42), the result (7.43) and the definition $\phi_0 \sim h^{1/\delta}$, one gets:

$$\psi_c \sim \frac{1}{h} \sim h^{1/\delta} h^{-\frac{2D}{D+2-\eta}} \tag{7.44}$$

Simple algebra gives the relation (7.30).

In actual calculations one usually obtains the critical exponent of the correlation length ν, and that of the correlation function η, while the rest of the exponents are derived from the relations (7.29)–(7.33) automatically.

7.3 Scaling

The concepts of the critical exponents and the correlation length are crucial for the theory of the second-order phase transitions. In the scaling theory of the critical phenomena it is implied that R_c is the only relevant spatial scale that exists in the system near T_c. As we have seen in the Ginzburg–Landau mean-field approach discussed above, at scales smaller than R_c all the spatial correlations are power like, which means that at scales much smaller than the correlation length everything must be scale invariant. On the other hand, at the phase-transition point the correlation length is infinite. Therefore, the properties of the system at scales smaller then R_c must be equivalent to those of the whole system at the phase-transition point.

The other important consequence of scale invariance is that the microscopic details of a system (lattice structure, etc.) must be irrelevant for the critical behavior. Critical properties are defined only by the 'global' characteristics of the system, such as space dimensionality, symmetry of the order parameter, etc. All the above arguments make a basis for the so-called *scaling hypothesis*, which claims that the macroscopic properties of a system at the critical point do not change after a global change of the spatial scale.

Let us consider the general consequences of this hypothesis. Let the Hamiltonian of a system be defined as follows:

$$H = \int d^D x \left[\frac{1}{2} (\nabla\phi(x))^2 + \sum_{n=1} h_n \phi^n(x) \right] \tag{7.45}$$

Here the set of parameters h_n describes a concrete system under consideration. In particular: $h_1 \equiv -h$ is the external field; $h_2 \equiv \tau$ is the 'mass' in the Ginzburg–Landau theory; $h_4 \equiv \frac{1}{4}g$; and the rest of the parameters describe other types of interaction. After the scale transformation:

$$x \to \lambda x \quad (\lambda > 1) \tag{7.46}$$

one gets:

$$\begin{aligned} \frac{1}{2}\int d^D x (\nabla\phi(x))^2 &\to \frac{1}{2}\lambda^{D-2} \int d^D x\, (\nabla\phi(\lambda x))^2 \\ h_n \int d^D x \phi^n(x) &\to \lambda^D h_n \int d^D x \phi^n(\lambda x) \end{aligned} \tag{7.47}$$

To leave the gradient term of the Hamiltonian (which is responsible for the scaling of the correlation functions) unchanged, one has to rescale the fields:

$$\phi(\lambda x) \to \lambda^{-\Delta_\phi} \phi(x) \tag{7.48}$$

with

$$\Delta_\phi = \frac{D-2}{2} \tag{7.49}$$

The *scale dimensions* Δ_ϕ define the critical exponent of the correlation function:

$$G(x) = \langle \phi(0)\phi(x) \rangle \sim |x|^{-2\Delta_\phi} \tag{7.50}$$

To leave the Hamiltonian (7.45) unchanged after these transformations one must also rescale the parameters h_n:

$$h_n \to \lambda^{-\Delta_n} h_n \tag{7.51}$$

where

$$\Delta_n = \frac{1}{2}(2-n)D + n \tag{7.52}$$

The quantities Δ_n are called the *scale dimensions* of the corresponding parameters h_n. In particular:

$$\Delta_1 \equiv \Delta_h = \frac{1}{2}D + 1 \tag{7.53}$$

$$\Delta_2 \equiv \Delta_\tau = 2 \tag{7.54}$$

$$\Delta_4 \equiv \Delta_g = 4 - D \tag{7.55}$$

Correspondingly, the rescaled parameters h_λ, τ_λ and g_λ of the Ginzburg–Landau Hamiltonian are:

$$h_\lambda = \lambda^{\Delta_h} h \tag{7.56}$$

$$\tau_\lambda = \lambda^{\Delta_\tau} \tau \tag{7.57}$$

$$g_\lambda = \lambda^{\Delta_g} g \tag{7.58}$$

These equations demonstrate the following:

(1) If the initial value of the 'mass' τ is non-zero, then the scale transformation makes the value of the rescaled τ_λ grow, and at the scale

$$\lambda_c \equiv R_c = |\tau|^{-\frac{1}{\Delta_\tau}} \tag{7.59}$$

the value of τ_λ becomes of the order of 1. This indicates that at $\lambda > R_c$ we are getting out of the scaling region, and the value R_c must be called the correlation length. Moreover, according to Eq. (7.59) for the critical exponent of the correlation length we find:

$$\nu = \frac{1}{\Delta_\tau} \tag{7.60}$$

(2) The value (and the critical exponent) of the critical field $h_c(\tau)$ can be obtained from the Eqs. (7.53) and (7.56), along the same lines:

$$h_{\lambda=R_c} = R_c^{\Delta_h} h_c \sim 1 \Rightarrow h_c \sim R_c^{-\Delta_h} \sim |\tau|^{\frac{\Delta_h}{\Delta_\tau}} \tag{7.61}$$

(3) If the dimension of the system is greater than 4, then according to Eqs. (7.55) and (7.58), $\Delta_g < 0$, and the rescaled value of the parameter g_λ tends to zero at infinite scales. Therefore, the theory becomes asymptotically Gaussian in this case. That is why the systems with dimensions $D > 4$ are described correctly by the Ginzburg–Landau theory.

On the other hand, at dimension $D < 4$, $\Delta_g > 0$, and the rescaled value of g_λ grows as the scale increases. In this case the situation becomes highly non-trivial because the asymptotic (infinite scale) theory becomes non-Gaussian. Nevertheless, if the dimension D is formally taken to be close to 4, such that the value of $\epsilon = 4 - D$ is treated as the small parameter, then the deviation from the Gaussian theory is also small in ϵ, and this allows us to treat such systems in terms of the perturbation theory (see next section). In the lucky case, if for some reason the series in ϵ appears to be 'good' and quickly converging, then one could hope to get the critical exponents close to the real ones if we set $\epsilon = 1$ in the final results.

It is a miracle, but although the actual series in ϵ appeares to be not 'good' at all (formally it is not even converging), the critical exponents obtained from the first few terms of the series taken at $\epsilon = 1$ ($D = 3$) turn out to be very close to the real ones.

7.4 Renormalization-group approach and ϵ-expansion

Let us assume that at large scales the asymptotic theory is described by the Hamiltonian (7.13) (for simplicity the external field h is taken to be zero):

$$H = \int d^D x \left[\frac{1}{2}(\nabla\phi(x))^2 + \frac{1}{2}\tau\phi^2(x) + \frac{1}{4}g\phi^4(x)\right] \tag{7.62}$$

where the field $\phi(x)$ is supposed to be slow-varying in space, such that the Fourier-transformed field $\phi(k)$:

$$\phi(x) = \int_{|k|<k_0} \frac{d^D k}{(2\pi)^D}\phi(k)\exp(ikx) \tag{7.63}$$

has only long-wave components: $| k |< k_0 \ll 1$. The parameters of the Hamiltonian are also assumed to be small: $|\tau| \ll 1$; $g \ll 1$. Correspondingly, the Fourier-transformed Hamiltonian is:

$$\begin{aligned} H_{k_0} = {} & \frac{1}{2}\int_{|k|<k_0} \frac{d^D k}{(2\pi)^D} k^2 \mid \phi(k) \mid^2 + \frac{1}{2}\tau \int_{|k|<k_0} \frac{d^D k}{(2\pi)^D} \mid \phi(k) \mid^2 + \frac{1}{4}g \\ & \times \int_{|k|<k_0} \frac{d^D k_1 d^D k_2 d^D k_3 d^D k_4}{(2\pi)^{4D}} \phi(k_1)\phi(k_2)\phi(k_3)\phi(k_4)\delta(k_1+k_2+k_3+k_4) \end{aligned} \tag{7.64}$$

In the most general terms the problem is to calculate the partition function:

$$Z = \left[\prod_{k=0}^{k_0} \int d\phi(k)\right] \exp\{-H_{k_0}(\phi)\} \tag{7.65}$$

and the corresponding free energy: $F = -\ln(Z)$.

The idea of the renormalization-group (RG) approach is described below.

In the **first step** one integrates only over the components of the field $\phi(k)$ in the limited wave band $\lambda k_0 < k < k_0$, where $\lambda \ll 1$. In the result we get a new Hamiltonian, which depends on the new cutoff λk_0:

$$\exp\{-\tilde{H}_{\lambda k_0}[\phi]\} \equiv \left[\prod_{k=\lambda k_0}^{k_0} \int d\phi(k)\right] \exp\left(-H_{k_0}[\phi]\right) \tag{7.66}$$

It is expected that under certain conditions the new Hamiltonian $\tilde{H}_{\lambda k_0}[\phi]$ has a structure similar to the original one, given by Eq. (7.64):

$$\begin{aligned}\tilde{H}_{\lambda k_0} &= \frac{1}{2}\tilde{a}(\lambda)\int_{|k|<\lambda k_0} \frac{d^Dk}{(2\pi)^D}k^2 \mid \phi(k) \mid^2 + \frac{1}{2}\tilde{\tau}(\lambda)\int_{|k|<\lambda k_0} \frac{d^Dk}{(2\pi)^D} \mid \phi(k) \mid^2 \\ &+ \frac{1}{4}\tilde{g}(\lambda)\int_{|k|<\lambda k_0} \frac{d^Dk_1 d^Dk_2 d^Dk_3 d^Dk_4}{(2\pi)^{4D}} \\ &\times \phi(k_1)\phi(k_2)\phi(k_3)\phi(k_4)\delta(k_1+k_2+k_3+k_4) + (\ldots)\end{aligned} \tag{7.67}$$

All additional terms that could appear in $\tilde{H}_{\lambda k_0}[\phi]$ after the integration (7.66) (denoted by '(...)') will be shown to be irrelevant for $\tau \ll 1$, $g \ll 1$, $\lambda \ll 1$, and $\epsilon = (4 - D) \ll 1$. In fact, the leading terms in (7.67) will be shown to be large in the parameter $\xi \equiv \ln(1/\lambda) \gg 1$, with the condition that $\epsilon \ln(1/\lambda) \ll 1$.

In the **second step** one makes the inverse scaling transformation (see Section 7.3) with the aim of restoring the original cutoff scale k_0:

$$\begin{aligned} k &\to \lambda k \\ \phi(\lambda k) &\to \theta(\lambda)\phi(k) \end{aligned} \tag{7.68}$$

The parameter $\theta(\lambda)$ should be chosen such that the coefficient of the $k^2 \mid \phi(k) \mid^2$ term remains the same as in the original Hamiltonian (7.64):

$$\theta = \lambda^{-\frac{D+2}{2}} (\tilde{a}(\lambda))^{-1/2} \tag{7.69}$$

The two steps given above compose the so-called *renormalization transformation*. The renormalized Hamiltonian is:

$$\begin{aligned} H_{k_0}^{(R)} &= \frac{1}{2}\int_{|k|<k_0} \frac{d^Dk}{(2\pi)^D}k^2 \mid \phi(k) \mid^2 + \frac{1}{2}\tau^{(R)}(\lambda)\int_{|k|<k_0} \frac{d^Dk}{(2\pi)^D} \mid \phi(k) \mid^2 \\ &+ \frac{1}{4}g^{(R)}(\lambda)\int_{|k|<k_0} \frac{d^Dk_1 d^Dk_2 d^Dk_3 d^Dk_4}{(2\pi)^{4D}} \\ &\times \phi(k_1)\phi(k_2)\phi(k_3)\phi(k_4)\delta(k_1+k_2+k_3+k_4) \end{aligned} \tag{7.70}$$

This Hamiltonian depends on the original cutoff k_0 whereas its parameters are renormalized:

$$\tau^{(R)}(\lambda) = \lambda^{-2}\,[\tilde{a}(\lambda)]^{-1}\,\tilde{\tau}(\lambda) \tag{7.71}$$

$$g^{(R)}(\lambda) = \lambda^{-(4-D)}\,[\tilde{a}(\lambda)]^{-2}\,\tilde{g}(\lambda) \tag{7.72}$$

The above RG transformation must be applied (infinitely) many times, and then the problem is to study the limiting properties of the renormalized Hamiltonian, which is expected to describe the asymptotic (infinite scale) properties of the system. In particular, it is hoped that the asymptotic Hamiltonian arrives at some fixed point Hamiltonian H^*, which is invariant with respect to the above RG transformation. The hypothesis about the existence of the fixed point (non-Gaussian) Hamiltonian H^*, which is invariant with respect to the scale transformations in the critical point, is nothing but a more 'mathematical' formulation of the scaling hypothesis discussed in Section 7.3.

Let us consider the RG procedure in more detail. To get the RG (Eqs. (7.71)–(7.72)) in explicit form one has to obtain the parameters $\tilde{a}(\lambda), \tilde{\tau}(\lambda), \tilde{g}(\lambda)$ by integrating over 'fast' degrees of freedom in Eq. (7.66). Let us separate the 'fast' fields (with $\lambda k_0 < |k| < k_0$) and the 'slow' fields (with $|k| < \lambda k_0$) explicitly:

$$\phi(x) = \tilde{\phi}(x) + \varphi(x);$$

$$\tilde{\phi}(x) = \int_{|k|<\lambda k_0} \frac{d^D k}{(2\pi)^D} \tilde{\phi}(k) \exp(ikx);$$

$$\varphi(x) = \int_{\lambda k_0<|k|<k_0} \frac{d^D k}{(2\pi)^D} \varphi(k) \exp(ikx) \tag{7.73}$$

Then the Hamiltonian (7.64) can be represented as follows:

$$H_{k_0}[\tilde{\phi}, \varphi] = H_{\lambda k_0}[\tilde{\phi}] + \frac{1}{2} \int_{\lambda k_0<|k|<k_0} \frac{d^D k}{(2\pi)^D} G_0^{-1}(k) \mid \varphi(k) \mid^2 + V[\tilde{\phi}, \varphi] \tag{7.74}$$

where

$$G_0(k) = k^{-2} \tag{7.75}$$

and

$$V[\tilde{\phi}, \varphi] = \frac{1}{2}\tau \int_{\lambda k_0<|k|<k_0} \frac{d^D k}{(2\pi)^D} \mid \varphi(k) \mid^2$$
$$+ \frac{3}{2} g \int \frac{d^D k_1 d^D k_2 d^D k_3 d^D k_4}{(2\pi)^{4D}} \tilde{\phi}(k_1)\tilde{\phi}(k_2)\varphi(k_3)\varphi(k_4)\delta(k_1+k_2+k_3+k_4)$$
$$+ g \int \frac{d^D k_1 d^D k_2 d^D k_3 d^D k_4}{(2\pi)^{4D}} \tilde{\phi}(k_1)\varphi(k_2)\varphi(k_3)\varphi(k_4)\delta(k_1+k_2+k_3+k_4)$$

Fig. 7.1. Diagrammatic representation of the interaction energy $V[\tilde{\phi}, \varphi]$.

$$+g \int \frac{d^D k_1 d^D k_2 d^D k_3 d^D k_4}{(2\pi)^{4D}} \tilde{\phi}(k_1)\tilde{\phi}(k_2)\tilde{\phi}(k_3)\varphi(k_4)\delta(k_1 + k_2 + k_3 + k_4)$$
$$+\frac{1}{4} g \int \frac{d^D k_1 d^D k_2 d^D k_3 d^D k_4}{(2\pi)^{4D}} \varphi(k_1)\varphi(k_2)\varphi(k_3)\varphi(k_4)\delta(k_1 + k_2 + k_3 + k_4) \quad (7.76)$$

In standard diagram notation the interaction term $V[\tilde{\phi}, \varphi]$ is shown in Fig. 7.1, where the wavy lines represent the 'slow' fields $\tilde{\phi}$, the straight lines represent the 'fast' fields φ, the solid circle represents the 'mass' τ, the open circle represents the interaction vertex g, and at each vertex the sum of entering 'impulses' k is zero.

Then, integration over the φs (Eq. (7.66)), yields:

$$\exp\{-\tilde{H}_{\lambda k_0}[\tilde{\phi}]\} = \exp\{-H_{\lambda k_0}[\tilde{\phi}]\}\langle\exp\{-V[\tilde{\phi}, \varphi]\}\rangle \quad (7.77)$$

where the averaging $\langle(\ldots)\rangle$ is performed as follows:

$$\langle(\ldots)\rangle \equiv \left[\prod_{k=\lambda k_0}^{k=k_0} \int d\varphi(k)\right] \exp\left(-\frac{1}{2}\int_{\lambda k_0<|k|<k_0} \frac{d^D k}{(2\pi)^D} G_0^{-1}(k) \mid \varphi(k) \mid^2\right) (\ldots) \quad (7.78)$$

Standard perturbation expansion in V gives:

$$\tilde{H}_{\lambda k_0}[\tilde{\phi}] = H_{\lambda k_0}[\tilde{\phi}] + \langle V\rangle - \frac{1}{2}\left[\langle V^2\rangle - \langle V\rangle^2\right] + \ldots \quad (7.79)$$

In terms of the diagrams (Fig. 7.1), the averaging $\langle\ldots\rangle$ is just the pairing of the straight lines. The non-zero contributions to $\langle V\rangle$ are shown in Fig. 7.2, where each closed loop is:

$$\int_{\lambda k_0<|k|<k_0} \frac{d^D k}{(2\pi)^D} G_0(k) = \frac{S_D}{(2\pi)^D} \int_{\lambda k_0}^{k_0} dk\, k^{(D-1)} \frac{1}{k^2}$$
$$= \frac{S_D}{(2\pi)^D (D-2)} k_0^{(D-2)} \left(1 - \lambda^{(D-2)}\right) \quad (7.80)$$

(here S_D is the surface area of a unite D-dimensional sphere).

In what follows we are going to study the limit case of the small cutoff k_0 (large spatial scales). Besides, at each RG step the rescaling parameter λ will also be assumed to be small, such that in all the integrations over the 'internal' ks ($\lambda k_0 < |k| < k_0$) the 'external' ks ($|k| < \lambda k_0$) can be considered as negligibly small.

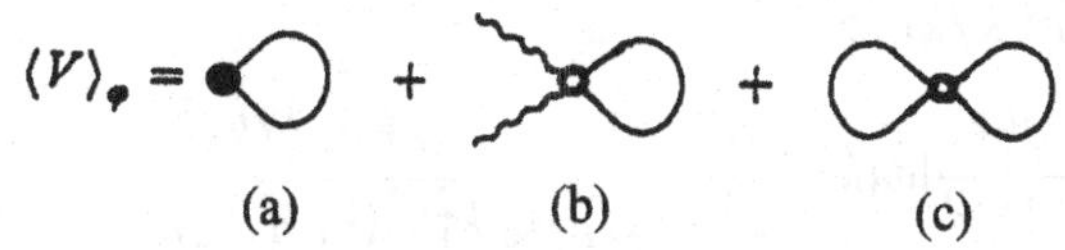

$\langle V\rangle_\varphi =$ (a) + (b) + (c)

Fig. 7.2. Diagrammatic representation of the first-order perturbation contribution $\langle V\rangle$.

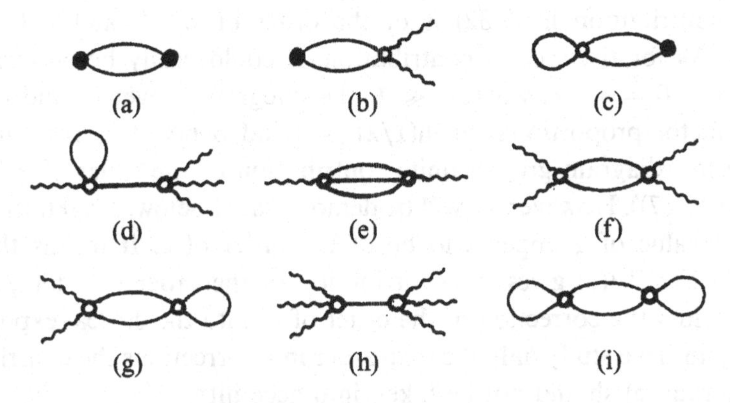

Fig. 7.3. Diagrammatic representation of the second-order perturbation contribution $\langle\langle V^2\rangle\rangle$.

The result for the first-order perturbation expansion $\langle V\rangle$ consists of three contributions. Diagrams (a) and (c) in Fig. 7.2 produce only irrelevant constants (they do not depend on $\tilde{\phi}$). Diagram (b) is proportional to $|\tilde{\phi}(k)|^2$ and gives the contribution to the mass term, but as this contribution is proportional to $k_0^{(D-2)}$, in the asymptotic region $k_0 \to 0$ it can be ignored as well. In fact we are going to look for the contributions, that: (1) do not depend on the value of the cutoff k_0; and (2) are large in the RG parameter $\xi \equiv \ln(1/\lambda) \gg 1$.

Consider the second-order perturbation contribution $\langle\langle V^2\rangle\rangle \equiv \langle V^2\rangle - \langle V\rangle^2$; see Fig. 7.3. Here the diagrams (a), (c) and (i) give irrelevant constants. Diagrams (d), (g) and (h) are proportional to the positive power of the cutoff k_0 and therefore their contribution is small.

The relevant diagrams are (b), (e) and (f). Diagram (e) is proportional to:

$$g^2 \int_{|k|<\lambda k_0} \frac{d^D k}{(2\pi)^D} |\tilde{\phi}(k)|^2 \int_{\lambda k_0<|k_{1,2}|<k_0} d^D k_1 d^D k_2 G_0(k_1) G_0(k_2) G_0(k+k_1+k_2)$$

$$= g^2 \int_{|k|<\lambda k_0} \frac{d^D k}{(2\pi)^D} |\tilde{\phi}(k)|^2 \int_{\lambda k_0<|k_{1,2}|<k_0} \frac{d^D k_1 d^D k_2}{k_1^2 k_2^2 (k+k_1+k_2)^2} \qquad (7.81)$$

Because $k \ll k_{1,2}$ the leading contribution in (7.81) is given by the first terms of the expansion in $k/k_{1,2}$:

$$g^2 \int_{|k|<\lambda k_0} \frac{d^D k}{(2\pi)^D} |\tilde{\phi}(k)|^2 \int_{\lambda k_0<|k_{1,2}|<k_0} \frac{d^D k_1 d^D k_2}{k_1^2 k_2^2 (k_1+k_2)^2}$$
$$+3g^2 \int_{|k|<\lambda k_0} \frac{d^D k}{(2\pi)^D} |\tilde{\phi}(k)|^2 k^2 \int_{\lambda k_0<|k_{1,2}|<k_0} \frac{d^D k_1 d^D k_2}{k_1^2 k_2^2 (k_1+k_2)^4} \tag{7.82}$$

The first contribution in (7.82) is of the order of $k_0^{(D-2)}$ and is therefore irrelevant. As for the second contribution, it could easily be checked that at dimension $D = 4 - \epsilon$, where $\epsilon \ll 1$, the integration over k_1 and k_2 does yield the factor proportional to $\ln(1/\lambda) \gg 1$ independent of the cutoff k_0. Therefore this diagram gives a finite contribution of the order of $g^2 \ln(1/\lambda)$ into $\tilde{a}$, (Eq. (7.67)). However, as will be demonstrated below, the renormalized fixed-point value of g appears to be of the order of ϵ. It means that the diagram in Fig. 7.3(e) gives the contribution of the order of $\epsilon^2 \ln(1/\lambda)$ in $\tilde{a}$ (which provides the correction of the order of ϵ^2 into the critical exponents). Therefore, until we study only the first-order in ϵ corrections the contribution of the diagram (e) should not be taken into account:

$$\tilde{a} = 1 + O(g^2)\xi \tag{7.83}$$

where $\xi \equiv \ln(1/\lambda)$.

Diagram (b) of Fig. 7.3 gives the following contribution:

$$\frac{3}{2} g\tau \int_{\lambda k_0<|k|<k_0} \frac{d^D k}{(2\pi)^D} \frac{1}{k^4} \int_{|k|<\lambda k_0} \frac{d^D k}{(2\pi)^D} |\tilde{\phi}(k)|^2$$
$$= \frac{3}{2} g\tau \frac{S_D}{(2\pi)^D} \frac{k_0^{(D-4)}(1-\lambda^{(D-4)})}{D-4} \int_{|k|<\lambda k_0} \frac{d^D k}{(2\pi)^D} |\tilde{\phi}(k)|^2 \tag{7.84}$$

For $D = 4 - \epsilon$, where $\epsilon \ll 1$, this gives the finite contribution to the parameter $\tilde{\tau}$:

$$\tilde{\tau} = \tau - \frac{3}{8\pi^2} \tau g \xi \tag{7.85}$$

(we have taken $S_{D=4} = 2\pi^2$)

For diagram (f) of Fig. 7.3 one gets:

$$\frac{9}{4} g^2 \int_{\lambda k_0<|k|<k_0} \frac{d^D k}{(2\pi)^D} \frac{1}{k^4} \int_{|k|<\lambda k_0} \frac{d^D k_1 d^D k_2 d^D k_3 d^D k_4}{(2\pi)^{4D}} \tilde{\phi}(k_1)\tilde{\phi}(k_2)\tilde{\phi}(k_3)\tilde{\phi}(k_4)$$
$$= \frac{9}{4} g^2 \frac{S_D}{(2\pi)^D} \frac{k_0^{(D-4)}(1-\lambda^{(D-4)})}{D-4}$$
$$\times \int_{|k|<\lambda k_0} \frac{d^D k_1 d^D k_2 d^D k_3 d^D k_4}{(2\pi)^{4D}} \tilde{\phi}(k_1)\tilde{\phi}(k_2)\tilde{\phi}(k_3)\tilde{\phi}(k_4) \tag{7.86}$$

For $D = 4 - \epsilon$ this gives the following contribution to the parameter $\tilde{g}$:

$$\tilde{g} = g - \frac{9}{8\pi^2} g^2 \xi \tag{7.87}$$

After the operation of rescaling to the original cutoff k_0, according to Eqs. (7.71)–(7.72) for the renormalized parameters $\tau^{(R)}$ and $g^{(R)}$, we get:

$$\begin{aligned} \tau^{(R)} &= \left(\tau - \frac{3}{8\pi^2} \tau g \xi\right) \exp(2\xi) \\ g^{(R)} &= \left(g - \frac{9}{8\pi^2} g^2 \xi\right) \exp(\epsilon\xi) \end{aligned} \tag{7.88}$$

For $g\xi \ll 1$ and $\epsilon\xi \ll 1$, these equations can be written as follows:

$$\begin{aligned} \ln(\tau^{(R)}) - \ln(\tau) &= 2\xi - \frac{3}{8\pi^2} g\xi \\ g^{(R)} - g &= \epsilon g \xi - \frac{9}{8\pi^2} g^2 \xi \end{aligned} \tag{7.89}$$

Assuming that the RG procedure is performed continuously, the evolution (as the scale changes) of the renormalized parameters can be described in terms of the differential equations. From Eqs. (7.89) one obtains:

$$\frac{d \ln|\tau|}{d\xi} = 2 - \frac{3}{8\pi^2} g \tag{7.90}$$

$$\frac{dg}{d\xi} = \epsilon g - \frac{9}{8\pi^2} g^2 \tag{7.91}$$

The fixed point solution g^* is defined by the condition $dg/d\xi = 0$, which yields:

$$g^* = \frac{8\pi^2}{9} \epsilon \tag{7.92}$$

Then, from Eq. (7.90) for the scale dimension Δ_τ one finds:

$$\Delta_\tau = 2 - \frac{1}{3} \epsilon \tag{7.93}$$

Correspondingly, according to Eq. (7.60) for the critical exponent ν, we obtain:

$$\tau(\xi) = \tau_0 \exp(\Delta_\tau \xi)\,; \qquad \nu = \frac{1}{2} + \frac{1}{12}\epsilon \tag{7.94}$$

Because the fixed-point value g^* is of the order of ϵ, according to Eqs. (7.83), (7.68) and (7.69) there are no corrections in the first order in ϵ to the scale dimensions Δ_ϕ of the field ϕ. Accordingly (see Eqs. (7.49) and (7.50)), in the first order in ϵ the critical exponent η, (Eq. (7.28)), of the correlation function $\langle\phi(0)\phi(x)\rangle$ remains zero, as in the Ginzburg–Landau theory.

Using relations (7.29)–(7.33), one can now easily find all the others critical exponents:

$$\alpha = \tfrac{1}{6}\epsilon \qquad \gamma = 1 + \tfrac{1}{6}\epsilon \quad \beta = \tfrac{1}{2} - \tfrac{1}{6}\epsilon$$
$$\delta = 3 + \epsilon \quad \mu = \tfrac{1}{3} \tag{7.95}$$

Below we give the values of the critical exponents in the first order in ϵ formally continued for the dimension $D = 3$ ($\epsilon = 1$). These are compared with the corresponding values given by numerical simulations and the Ginzburg–Landau theory.

		ϵ-expansion	Numerical simulations	Ginzburg–Landau
Order parameter:	β	0.333	0.312 ± 0.003	0.5
	δ	4	5.15 ± 0.02	3
Specific heat:	α	0.167	0.125 ± 0.015	0
Susceptibility:	γ	1.167	1.250 ± 0.003	1
Correlation length:	ν	0.583	0.642 ± 0.003	0.5
Correlation function:	η	0	0.055 ± 0.010	0

(7.96)

It is interesting to note that although the RG ϵ-expansion procedure described above is mathematically not well grounded, it provides rather accurate values for the critical exponents. For obtaining results in the second order in ϵ one proceeds in a similar way, taking into account the next order in the ϵ diagrams (see e.g. [10]).

7.5 Specific heat singularity in four dimensions

Although in dimensions $D = 4$ the critical exponent α is zero, it does not necessarily mean that the specific heat is not singular at the critical point. Actually in this case the specific heat is logarithmically (and not power-law) divergent. As a matter of useful exercise, let us calculate the specific heat singularity for the four dimensions.

According to the definition of the specific heat (see Eqs. (7.34) and (7.35)) we have:

$$C = -T\frac{\partial^2 f}{\partial T^2} = \frac{1}{V}\int d^4x \int d^4x' \langle\langle\phi^2(x)\phi^2(x')\rangle\rangle \simeq \int_{|x|<R_c(\tau)} d^4x \langle\langle\phi^2(0)\phi^2(x)\rangle\rangle \tag{7.97}$$

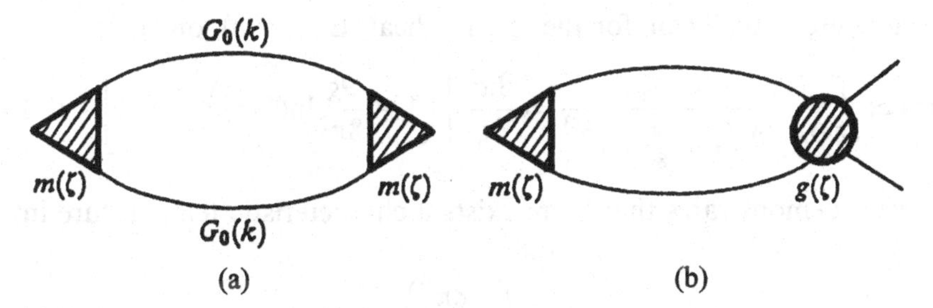

Fig. 7.4. (a) Diagrammatic representation of the specific heat. (b) The diagram that contributes to the renormalization of the 'dressed' mass $m(\xi)$.

Here the upper cutoff in the spatial integration is taken to be the correlation length, $R_c(\tau) \sim |\tau|^{-1/2}$, because at larger scales all the correlations decay exponentially. The integral in Eq. (7.97) can be calculated by summing up the so-called 'parquette' diagrams [31] shown in Fig. 7.4. The idea of the 'parquette' calculations is that all the contributions from the ϕ^4 interactions in the correlation function $\langle\langle\phi^2(x)\phi^2(x')\rangle\rangle$ can be collected into the mass-like vertex $m(\xi)$:

$$C \simeq \int_{|k|>\sqrt{\tau}} \frac{d^4k}{(2\pi)^4} G_0^2(k) \left(\frac{m(k)}{\tau}\right)^2 \sim \int_{|k|>\sqrt{\tau}} \frac{dk}{k} \left(\frac{m(k)}{\tau}\right)^2 \sim \int_{\xi<\ln(1/\tau)} d\xi \left(\frac{m(\xi)}{\tau}\right)^2 \tag{7.98}$$

Here the renormalization of the 'dressed' mass $m(\xi)$ is defined by the diagram shown in Fig. 7.4(b) (see also Eqs. (7.84)–(7.87)):

$$m^{(R)} = m - 3mg \int_{\lambda k_0<|k|<k_0} \frac{d^Dk}{(2\pi)^D} G_0^2(k) \to m - \frac{3}{8\pi^2} mg\xi \tag{7.99}$$

where, as usual, $\xi \equiv \ln(1/\lambda)$. In differential form:

$$\frac{dm(\xi)}{d\xi} = -\frac{3}{8\pi^2} m(\xi) g(\xi) \tag{7.100}$$

with initial conditions: $m(\xi = 0) = \tau$. The renormalization of the interaction parameter $g(\xi)$ for the dimension $D = 4$ is defined by the RG equation (7.91) with $\epsilon = 0$:

$$\frac{dg(\xi)}{d\xi} = -\frac{9}{8\pi^2} g^2(\xi) \tag{7.101}$$

The solution of Eqs. (7.100)–(7.101) is:

$$\begin{aligned} m(\xi) &= \tau \left(1 + \frac{9g}{8\pi^2}\xi\right)^{-1/3} \\ g(\xi) &= g \left(1 + \frac{9g}{8\pi^2}\xi\right)^{-1} \end{aligned} \tag{7.102}$$

where $g \equiv g(\xi = 0)$. Then, for the specific heat, Eq. (7.98), one gets:

$$C(\tau) \simeq \int_{\xi<\ln(1/\tau)} \frac{d\xi}{\left(1+\frac{9g}{8\pi^2}\xi\right)^{2/3}} = \frac{8\pi^2}{3g}\left[\left(1+\frac{9g}{8\pi^2}\ln(1/\tau)\right)^{1/3} - 1\right] \quad (7.103)$$

This result demonstrates that there exists a characteristic temperature interval:

$$\tau_g \sim \exp\left(-\frac{8\pi^2}{9g}\right) \ll 1 \quad (7.104)$$

such that at temperatures not too close to T_c, $\tau_g \ll |\tau| \ll 1$, the system is Gaussian (it does not depend on the non-Gaussian interaction parameter g):

$$C(\tau) \sim \ln(1/\tau) \quad (7.105)$$

This result could be easily obtained just in the framework of the Gaussian Ginzburg–Landau theory:

$$C(\tau) \sim \int d^4x\langle\langle\phi^2(0)\phi^2(x)\rangle\rangle \sim \int_{|k|<1} \frac{d^4k}{(k^2+\tau)^2} \sim \int_{\sqrt{\tau}}^{1} \frac{d^4k}{k^4} \sim \ln(1/\tau) \quad (7.106)$$

On the other hand, in the close vicinity of the critical point ($\tau \ll \tau_g$) the theory becomes non-Gaussian, and the result for the specific heat becomes less trivial:

$$C(\tau) \sim \frac{1}{g}\left[g\ln(1/\tau)\right]^{1/3} \quad (7.107)$$

Thus, although the critical exponent α is zero for the four-dimensional system, the specific heat still remains (logarithmically) divergent at the critical point.

8

Critical phenomena in systems with disorder

8.1 Harris criterion

In studies of the phase-transition phenomena, the systems considered before were assumed to be perfectly homogeneous. In real physical systems, however, some defects or impurities are always present. Therefore, it is natural to consider what effect impurities might have on the phase-transition phenomena. As we have seen in the previous chapter, the thermodynamics of the second-order phase transition is dominated by large-scale fluctuations. The dominant scale, or the correlation length, $R_c \sim |T/T_c - 1|^{-\nu}$ grows as T approaches the critical temperature T_c, where it becomes infinite. The large-scale fluctuations lead to singularities in the thermodynamical functions as $|\tau| \equiv |T/T_c - 1| \to 0$. These singularities are the main subject of the theory.

If the concentration of impurities is small, their effect on the critical behavior remains negligible so long as R_c is not too large, i.e. for T not too close to T_c. In this regime the critical behavior will be essentially the same as in the perfect system. However, as $|\tau| \to 0$ ($T \to T_c$) and R_c becomes larger than the average distance between impurities, their influence can become crucial.

As T_c is approached the following change of length scale takes place. First, the correlation length of the fluctuations becomes much larger than the lattice spacing, and the system 'forgets' about the lattice. The only relevant scale that remains in the system in this regime is the correlation length $R_c(\tau)$. When we move close to the critical point, R_c grows and becomes larger than the average distance between the impurities, so that the effective concentration of impurities, measured with respect to the correlation length, becomes large. It should be stressed that such a situation is reached for an arbitrary small initial concentration u. The value of u affects only the width of the temperature region near T_c in which the effective concentration becomes effectively large. If $uR_c^D \gg 1$ there are no grounds for believing that the effect of impurities will be small.

Originally, many years ago, it was generally believed that impurities either completely destroyed the long-range fluctuations, such that the singularities of the thermodynamical functions are smoothed out [32, 33], or could produce only a shift of a critical point but could not effect the critical behavior itself, so that the critical exponents remain the same as in the pure system [34]. Later it was realized that an intermediate situation is also possible, in which a new critical behavior, with new critical exponents, is established sufficiently close to the phase-transition point [35]. Moreover, a criterion, the so-called Harris criterion, has also been developed, which makes it possible to predict qualitatively the effect of impurities by using the critical exponents of the pure system only [33, 35]. According to this criterion, the impurities change the critical behavior only if the specific heat exponent α of the pure system is positive (the specific heat of the pure system is divergent in the critical point). In the opposite case, $\alpha < 0$ (the specific heat is finite), the impurities appear to be irrelevant, i.e. their presence does not affect the critical behavior.

Let us consider this point in more detail. It would be natural to assume that in the ϕ^4-Hamiltonian (Section 7.4) the presence of impurities manifests itself as small random spatial fluctuations of the reduced transition temperature τ. Then, near the phase-transition point, the D-dimensional Ising-like systems can be described in terms of the scalar field Ginzburg–Landau Hamiltonian with a double-well potential:

$$H = \int d^D x \left[\frac{1}{2} (\nabla\phi(x))^2 + \frac{1}{2} (\tau - \delta\tau(x))\, \phi^2(x) + \frac{1}{4} g\phi^4(x) \right] \tag{8.1}$$

Here the quenched disorder is described by random fluctuations of the effective transition temperature $\delta\tau(x)$ whose probability distribution is taken to be symmetric and Gaussian:

$$P[\delta\tau] = p_0 \exp\left(-\frac{1}{4u} \int d^D x (\delta\tau(x))^2 \right) \tag{8.2}$$

where $u \ll 1$ is the small parameter that describes the disorder, and p_0 is the normalization constant. For notational simplicity, we define the sign of $\delta\tau(x)$ in Eq. (8.1) so that positive fluctuations lead to locally ordered regions, whose effects are the object of our study.

Configurations of the fields $\phi(x)$ that correspond to local minima in H satisfy the saddle-point equation:

$$-\Delta\phi(x) + \tau\phi(x) + g\phi^3(x) = \delta\tau(x)\phi(x) \tag{8.3}$$

Such localized solutions exist in regions of space where $\tau - \delta\tau(x)$ assumes negative values. Clearly, the solutions of Eq. (8.3) depend on a particular configuration of the function $\delta\tau(x)$ being inhomogeneous. Let us estimate under which conditions the quenched fluctuations of the effective transition temperature are the dominant factor for the local minima field configurations.

Let us consider a large region Ω_L of a linear size $L \gg 1$. The spatially average value of the function $\delta\tau(x)$ in this region can be defined as follows:

$$\delta\tau(\Omega_L) = \frac{1}{L^D} \int_{x \in \Omega_L} d^D x \delta\tau(x) \tag{8.4}$$

Correspondingly, for the characteristic value of the temperature fluctuations (averaged over realizations) in this region we get:

$$\delta\tau_L = \sqrt{\overline{\delta\tau^2(\Omega_L)}} = \sqrt{2u} L^{-D/2} \tag{8.5}$$

Then, according to Eq. (8.3) the average value of the order parameter $\phi(\Omega_L)$ in this region can be estimated from the equation:

$$\tau + g\phi^2 = \delta\tau(\Omega_L) \tag{8.6}$$

One can easily see that if the value of τ is sufficiently small, i.e. if

$$\delta\tau(\Omega_L) \gg \tau \tag{8.7}$$

then the solutions of Eq. (8.6) are defined only by the value of the random temperature:

$$\phi(\Omega_L) \simeq \pm \left(\frac{\delta\tau(\Omega_L)}{g} \right)^{1/2} \tag{8.8}$$

Now let us estimate up to which sizes of locally ordered regions this may occur. According to Eq. (8.5) the condition $\delta\tau_L \gg \tau$ yields:

$$L \ll \frac{u^{1/D}}{\tau^{2/D}} \tag{8.9}$$

On the other hand, the estimation of the order parameter in terms of the saddle-point equation (8.6) can be correct only at scales much larger than the correlation length $R_c \sim \tau^{-\nu}$. Thus, one has the lower bound for L:

$$L \gg \tau^{-\nu} \tag{8.10}$$

Therefore, quenched temperature fluctuations are relevant when

$$\tau^{-\nu} \ll \frac{u^{1/D}}{\tau^{2/D}} \tag{8.11}$$

or

$$\tau^{2-\nu D} \ll u \tag{8.12}$$

According to the scaling relations, Eq. (2.55), one has $2 - \nu D = \alpha$. Thus one recovers the Harris criterion: if the specific heat critical exponent of the pure system is positive, then in the temperature interval

$$\tau < \tau_* \equiv u^{1/\alpha} \tag{8.13}$$

the disorder becomes relevant. This argument identifies $1/\alpha$ as the crossover exponent associated with randomness.

On the other hand, if the critical exponent $\alpha = 2 - \nu D < 0$, the condition (8.12) can not be satisfied near T_c (at $\tau \ll 1$), and therefore in this case a weak disorder remains irrelevant in the critical region.

A special consideration is required in the marginal situation $\alpha = 0$, which takes place, for instance, in the four-dimensional ϕ^4-model (Section 7.5), or in the two-dimensional Ising model (Chapter 5). It will be shown that although the specific heat exponent in the disordered models also remains zero, the actual forms of the logarithmic singularities are affected by the disorder.

8.2 Critical exponents in the ϕ^4-theory with disorder

To simplify further derivations, it turns out that instead of the Ising model (described by the scalar field Hamiltonian (8.1)) it is more convenient to consider the more general case of the *vector* ferromagnetic system described by the p-component order parameter $\phi_i(x)$ $(i = 1, 2, \ldots, p)$. In the vicinity of the critical point such systems are described by the following continuous Hamiltonian:

$$H[\delta\tau, \phi_i] = \int d^D x \left[\frac{1}{2} \sum_{i=1}^{p} (\nabla \phi_i(x))^2 + \frac{1}{2}(\tau - \delta\tau(x)) \times \sum_{i=1}^{p} \phi_i^2(x) + \frac{1}{4} g \sum_{i,j=1}^{p} \phi_i^2(x) \phi_j^2(x) \right] \tag{8.14}$$

where the random temperature $\delta\tau(x)$ is described by the Gaussian distribution (8.2). In particular, the Hamiltonian with $p = 1$ describes the Ising model, the one with $p = 2$ describes the XY-model, and the one with $p = 3$ describes the Heisenberg ferromagnet. It will be shown below that the Ising model constitutes a special (more difficult) case, and therefore to derive the qualitative phenomena discussed in this chapter it is more convenient to consider the systems with $p > 1$. Besides, as we will see below, the values and the signs of the specific heat critical exponents of the corresponding pure systems (with $\delta\tau(x) \equiv 0$) depend on the value of the parameter p (at $p < 4$, $\alpha > 0$, while at $p > 4$, $\alpha < 0$), and this makes it possible to verify by explicit calculations the validity of the Harris criterion discussed in the previous section.

In terms of the replica approach (Section 1.3) we have to calculate the following replica partition function:

$$Z_n = \overline{\left(\int D\phi_i(x) \exp\{-H[\delta\tau, \phi]\}\right)^n}$$
$$= \int D\delta\tau(x) \int D\phi_i^a(x) \exp\left[-\frac{1}{4u}\int d^D x (\delta\tau(x))^2 - \sum_{a=1}^{n} H[\delta\tau, \phi_i^a]\right] \quad (8.15)$$

where the superscript a labels the replicas. (Here and in what follows all irrelevant pre-exponential factors are omitted.) After Gaussian integration over $\delta\tau(x)$ one gets:

$$Z_n = \int D\phi_i^a(x) \exp\left(-H_n[\phi_i^a]\right) \quad (8.16)$$

where

$$H_n[\phi_i^a] = \int d^D x \left[\frac{1}{2}\sum_{i=1}^{p}\sum_{a=1}^{n} (\nabla\phi_i^a(x))^2 + \frac{1}{2}\tau \sum_{i=1}^{p}\sum_{a=1}^{n} (\phi_i^a(x))^2 \right.$$
$$\left. + \frac{1}{4}\sum_{i,j=1}^{p}\sum_{a,b=1}^{n} g_{ab}(\phi_i^a(x))^2(\phi_j^b(x))^2 \right] \quad (8.17)$$

and

$$g_{ab} = g\delta_{ab} - u \quad (8.18)$$

Now we shall calculate the critical exponents using the RG procedure developed in Section 7.4 for dimension $D = 4-\epsilon$ assuming that $\epsilon \ll 1$. Taking into account the vector and the replica components, the ϕ^4 interaction terms in the Hamiltonian (8.17) can be represented in terms of the diagram shown in Fig. 8.1. If we proceed similarly to the calculations of Section 7.4 we find that the (one-loop) renormalization of the interaction parameters g_{ab} (Fig. 8.1) are given by the diagrams shown in Fig. 8.2. Taking into account corresponding combinatoric factors one obtains the following contributions:

$$(a) \to g_{ab}^2 \int_{\lambda k_0 < |k| < k_0} \frac{d^D k}{(2\pi)^D} G_0^2(k)|_{\epsilon \ll 1} \simeq g_{ab}^2 \frac{1}{8\pi^2}\ln\left(\frac{1}{\lambda}\right)$$

$$(b) \to \frac{1}{2}(g_{aa} + g_{bb})g_{ab} \int_{\lambda k_0 < |k| < k_0} \frac{d^D k}{(2\pi)^D} G_0^2(k)|_{\epsilon \ll 1} \simeq \frac{1}{2}(g_{aa} + g_{bb})g_{ab} \frac{1}{8\pi^2} \ln\left(\frac{1}{\lambda}\right)$$

$$(c) \to \frac{p}{4} \sum_{c=1}^{n} g_{ac} g_{cb} \int_{\lambda k_0 < |k| < k_0} \frac{d^D k}{(2\pi)^D} G_0^2(k)|_{\epsilon \ll 1} \simeq \frac{1}{4} p \sum_{c=1}^{n} g_{ac} g_{cb} \frac{1}{8\pi^2} \ln\left(\frac{1}{\lambda}\right) \quad (8.19)$$

The corresponding RG equations are:

$$\frac{dg_{ab}}{d\xi} = \epsilon g_{ab} - \frac{1}{8\pi^2}\left[4g_{ab}^2 + 2(g_{aa} + g_{bb})g_{ab} + p\sum_{c=1}^{n} g_{ac} g_{cb}\right] \quad (8.20)$$

Taking into account the definition (8.18) one easily gets two RG equations for two interaction parameters $\tilde{g} \equiv g_{aa} = g - u$ and $g_{a \neq b} = -u$:

$$\frac{d\tilde{g}}{d\xi} = \epsilon \tilde{g} - \frac{1}{8\pi^2}\left[(8 + p)\tilde{g}^2 + p(n-1)u^2\right]$$

$$\frac{du}{d\xi} = \epsilon u - \frac{1}{8\pi^2}\left[(4 + 2p)\tilde{g}u - (4 + p(n-2))u^2\right] \quad (8.21)$$

In the limit $n \to 0$ we obtain:

$$\frac{d\tilde{g}}{d\xi} = \epsilon \tilde{g} - \frac{1}{8\pi^2}\left[(8 + p)\tilde{g}^2 - pu^2\right]$$

$$\frac{du}{d\xi} = \epsilon u - \frac{1}{8\pi^2}\left[(4 + 2p)\tilde{g}u - (4 - 2p)u^2\right] \quad (8.22)$$

Similarly, the renormalization of the 'mass' term $\tau(\phi_i^a(x))^2$ is given by the diagrams shown in Fig. 8.3. Their contributions are:

$$(a) \to \tau g_{aa} \int_{\lambda k_0 < |k| < k_0} \frac{d^D k}{(2\pi)^D} G_0^2(k)|_{\epsilon \ll 1} \simeq \tau g_{aa} \frac{1}{8\pi^2} \ln\left(\frac{1}{\lambda}\right)$$

$$(b) \to \frac{1}{2} p\tau \sum_{c=1}^{n} g_{ca} \int_{\lambda k_0 < |k| < k_0} \frac{d^D k}{(2\pi)^D} G_0^2(k)|_{\epsilon \ll 1} \simeq \frac{1}{2} p\tau \sum_{c=1}^{n} g_{ca} \frac{1}{8\pi^2} \ln\left(\frac{1}{\lambda}\right) \quad (8.23)$$

Note that the above contributions do not depend on the replica index a. The corresponding RG equation for the renormalized 'mass' τ is:

$$\frac{d(\ln|\tau|)}{d\xi} = 2 - \frac{1}{8\pi^2}\left[2g_{aa} + p\sum_{c=1}^{n} g_{ca}\right] \quad (8.24)$$

In the limit $n \to 0$ we finally obtain:

$$\frac{d(\ln|\tau|)}{d\xi} = 2 - \frac{1}{8\pi^2}\left[(2 + p)\tilde{g}(\xi) + pu(\xi)\right] \quad (8.25)$$

where the renormalized interaction parameters $\tilde{g}(\xi)$ and $u(\xi)$ are defined by Eqs. (8.22).

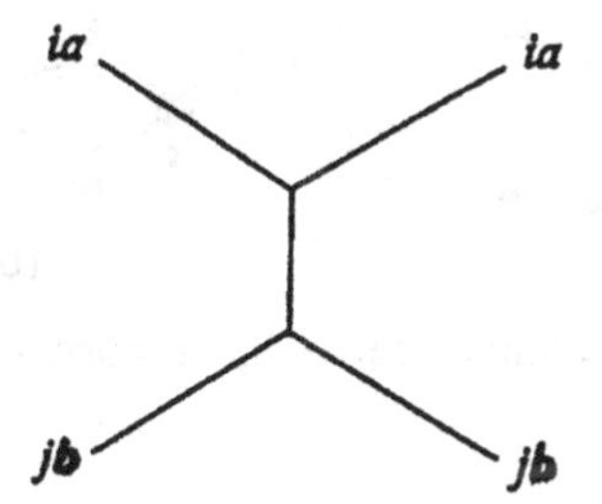

Fig. 8.1. Diagrammatic representation of the interaction term $g_{ab}(\phi_i^a(x))^2(\phi_j^b(x))^2$.

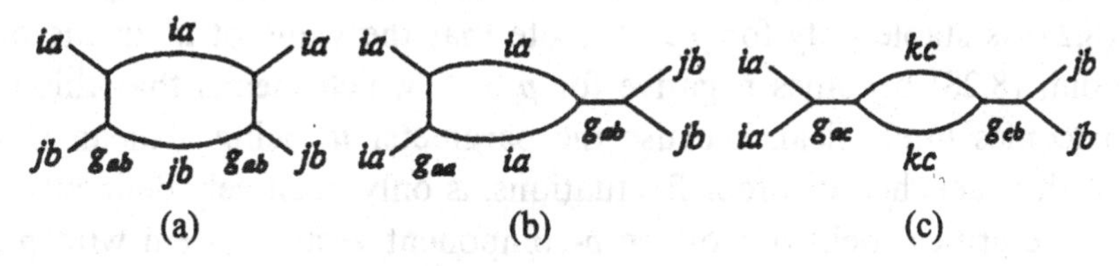

Fig. 8.2. The diagrams that contribute to the interaction term $g_{ab}(\phi_i^a(x))^2(\phi_j^b(x))^2$.

The fixed-point values $\tilde{g}^*$ and u^* are defined by the conditions $d\tilde{g}^*/d\xi = 0$, $du^*/d\xi = 0$, which according to Eqs. (8.22) yield:

$$(8+p)\tilde{g}^2 - pu^2 = 8\pi^2\epsilon g$$
$$(4+2p)\tilde{g}u - (4-2p)u^2 = 8\pi^2\epsilon u \tag{8.26}$$

These equations have two non-trivial solutions:

$$\tilde{g}^* = \frac{8\pi^2}{p+8}\epsilon; \qquad u^* = 0 \tag{8.27}$$

and

$$\tilde{g}^* = \epsilon\pi^2\frac{p}{2(p-1)}; \qquad u^* = \epsilon\pi^2\frac{4-p}{2(p-1)} \tag{8.28}$$

The first solution, Eq. (8.27), describes the pure system without disorder. Using Eq. (8.25) and the relations (7.60), (7.29) for the critical exponents of the pure system (we mark them by the label '(0)') one gets:

$$\Delta_\tau^{(0)} = 2 - \frac{1}{8\pi^2}(2+p)\tilde{g}^*_{(0)} = 2 - \frac{2+p}{8+p}\epsilon; \;\Rightarrow\; \nu_{(0)} = \frac{1}{\Delta_\tau^{(0)}} \simeq \frac{1}{2} + \frac{2+p}{4(8+p)}\epsilon \tag{8.29}$$

$$\alpha_{(0)} = 2 - (4-\epsilon)\nu_{(0)} \simeq \frac{4-p}{2(8+p)}\epsilon \tag{8.30}$$

and by using relations (7.29)–(7.33) the rest of the exponents are obtained automatically.

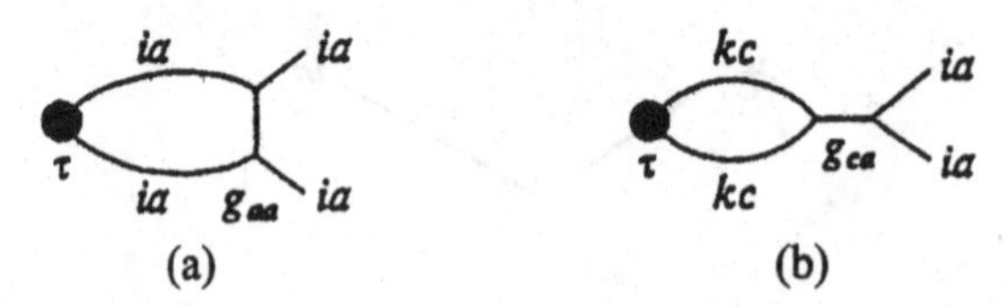

Fig. 8.3. The diagrams that contribute to the renormalization of the 'mass' term $\tau(\phi_i^a(x))^2$.

Simple linear analysis of the 'dynamics' defined by the RG equations (8.22) in the vicinity of the fixed points (8.27) and (8.28) shows that the 'pure' fixed point (8.27) is stable only for $p > 4$. Note that the value of u^* in the other fixed point (8.28) becomes negative for $p > 4$, which means that this fixed point becomes unphysical, because the parameter u, being a mean square value of the quenched disorder fluctuations, is only positively defined.

Thus, the critical behavior of the p-component vector system with $p > 4$ is not modified by the presence of quenched disorder. It should be stressed that it is just the case when the specific heat critical exponent α is negative, Eq. (8.30), in accordance with the Harris criterion (Section 8.1).

For $p < 4$ the 'pure' fixed point (8.27) becomes unstable and the critical properties of the system are defined by the 'random' fixed point given by Eq. (8.28). Using Eq. (8.25), one gets:

$$\Delta_\tau = 2 - \frac{1}{8\pi^2}[(2+p)\tilde{g}^* + pu^*] = 2 - \frac{3p}{8(p-1)}\epsilon; \;\Rightarrow\; \nu = \frac{1}{\Delta_\tau} \simeq \frac{1}{2} + \frac{3p}{32(p-1)}\epsilon \tag{8.31}$$

$$\alpha = 2 - (4-\epsilon)\nu \simeq -\frac{4-p}{8(p-1)}\epsilon \tag{8.32}$$

where p must be greater than 1. The rest of the exponents are obtained automatically.

Because at $p = 1$ the equations (8.22) become degenerate, the case of the one-component (Ising) system requires more detailed consideration. It turns out that such degeneracy is the property only of the first-order in ϵ approximation. It can be proved that taking into account next-order in ϵ diagrams the degeneracy of the RG equations is lifted. It can be shown then that a new 'random' fixed point of the RG equations exists for $p = 1$ as well, and in this case the corrections to the critical exponents appear to be of the order of $\sqrt{\epsilon}$ [35]. We omit this analysis here because it is technically much more cumbersome, while on a qualitative level it provides the results similar to those obtained above.

Thus, in agreement with the Harris criterion (Section 8.1) in the vector p-component system with $p < 4$ the critical behavior is modified by the presence of quenched disorder. In the vicinity of the critical point a new critical regime appears, and it is described by a new set of (universal) critical exponents. Note that the 'random' critical exponent of the specific heat (8.32) appears to be negative, unlike that of the pure system. Therefore, the disorder makes the specific heat finite (although still singular) at the critical point, unlike the divergent specific heat of the corresponding pure system.

Nevertheless, regardless of the apparent selfconsistency of the results obtained above, it should be stressed that owing to *non-perturbative* spin-glass-type phenomena the relevance to the real physics of the approach considered in this section may be questioned (see next chapter).

8.3 Critical behavior of the specific heat in four dimensions

In the full analogy with the corresponding considerations for the pure systems (Eqs. (7.97) and (7.98), Section 7.5) for the singular part of the specific heat at $D = 4$ we get:

$$C \simeq \int_{|k|>\sqrt{\tau}} \frac{d^4k}{(2\pi)^4} G_0^2(k) \left(\frac{m(k)}{\tau}\right)^2 \sim \int_{\xi<\ln(1/\tau)} d\xi \left(\frac{m(\xi)}{\tau}\right)^2 \tag{8.33}$$

Here the renormalization of the 'dressed' mass $m(\xi)$ is defined by the 'parquette' diagrams of Fig. 8.3. Accordingly, the renormalizations of the interaction parameters $\tilde{g}(\xi)$ and $u(\xi)$ are defined by the RG equations (8.22) with $\epsilon = 0$:

$$\frac{d(\ln|m|)}{d\xi} = -\frac{1}{8\pi^2}[(2+p)\tilde{g} + pu] \tag{8.34}$$

$$\frac{d\tilde{g}}{d\xi} = -\frac{1}{8\pi^2}\left[(8+p)\tilde{g}^2 - pu^2\right] \tag{8.35}$$

$$\frac{du}{d\xi} = -\frac{1}{8\pi^2}\left[(4+2p)\tilde{g}u - (4-2p)u^2\right] \tag{8.36}$$

The initial conditions are: $m(\xi = 0) = \tau$; $\tilde{g}(\xi = 0) = g_0$; $u(\xi = 0) = u_0$.

In the pure system $u = 0$, and the asymptotic solutions for $m(\xi)$ and $\tilde{g}(\xi) \equiv g(\xi)$ are:

$$g(\xi \to \infty) \sim \frac{8\pi^2}{8+p}\xi^{-1}$$

$$m(\xi \to \infty) \sim \xi^{-\frac{2+p}{8+p}} \tag{8.37}$$

Integration in the Eq. (8.33) yields the following specific heat singularity:

$$C \sim \left[\ln(\frac{1}{\tau})\right]^{\frac{4-p}{8+p}} \tag{8.38}$$

For the system with non-zero disorder interaction parameter u, one finds the following asymptotic solutions of Eqs. (8.34)–(8.36):

$$\tilde{g}(\xi \to \infty) \sim \pi^2 \frac{p}{2(p-1)} \xi^{-1};$$

$$u(\xi \to \infty) \sim \pi^2 \frac{(4-p)}{2(p-1)} \xi^{-1};$$

$$m(\xi \to \infty) \sim \xi^{-\frac{3p}{8(p-1)}} \tag{8.39}$$

Such solutions exist only for $1 < p < 4$. The case of one-component field, $p = 1$, requires special consideration. Here one has to take into account second-order loop terms, which make the analysis rather cumbersome, and we do not consider it here. On a qualitative level, however, the result for the specific heat appears to be similar to those for $1 < p < 4$: the one-component system with impurities exhibits a new type of (logarithmic) singularity.

On the other hand, one can easily check that at $p > 4$ the asymptotic solution of Eqs. (8.34)–(8.36) turns out to be the same as in the pure system, Eq. (8.37).

In the case $1 < p < 4$, the integration in the Eq. (8.33) yields:

$$C \sim \left[\ln\left(\frac{1}{\tau}\right)\right]^{-\frac{4-p}{4(p-1)}} \tag{8.40}$$

It is interesting to note that although at the dimension $D = 4$ the critical exponent α of the specific heat is zero, the Harris criterion, taken in the generalized form, still works. Namely, if the specific heat of the pure system is divergent at the critical point (the case of $p < 4$, Eq. (8.38)), the disorder appears to be relevant for the critical behavior, and changes the behavior of the specific heat into a new type of (universal) singularity (Eq. (8.40)). Otherwise, if the specific heat of the pure system is finite at the critical point ($p > 4$, Eq. (8.38)), then the presence of the disorder does not modify the critical behavior.

9

Spin-glass effects in the critical phenomena

9.1 Non-perturbative degrees of freedom

In this chapter we consider non-trivial spin-glass (SG) effects produced by weak quenched disorder, which have been ignored in the previous chapter. It will be shown that these effects could dramatically change the whole physical scenario of the critical phenomena.

According to the traditional point of view (considered in the previous chapter) the effects produced by weak quenched disorder in the critical region could be summarized as follows. If α, the specific heat exponent of the pure system, is greater than zero (i.e. the specific heat of the pure system is divergent at the critical point) the disorder is relevant for the critical behavior, and a new universal critical regime, with new critical exponents, is established sufficiently close to the phase-transition point $\tau \ll \tau_u \equiv u^{1/\alpha}$. In contrast, when $\alpha < 0$ (the specific heat is finite), the disorder appears to be irrelevant, i.e. their presence does not affect the critical behavior. Actually, if the disorder is relevant for the critical behavior, the situation could appear to be much more sophisticated. Let us consider the physical motivation of the traditional RG approach in more detail.

Near the phase-transition point the D-dimensional Ising-like systems can be described in terms of the scalar field Ginzburg–Landau Hamiltonian with a double-well potential:

$$H = \int d^D x \left[\frac{1}{2} (\nabla\phi(x))^2 + \frac{1}{2}(\tau - \delta\tau(x))\phi^2(x) + \frac{1}{4} g\phi^4(x) \right] \tag{9.1}$$

Here, as usual, the quenched disorder is described by random fluctuations of the effective transition temperature $\delta\tau(x)$ whose probability distribution is taken to be symmetric and Gaussian:

$$P[\delta\tau] = p_0 \exp\left[-\frac{1}{4u} \int d^D x (\delta\tau(x))^2 \right] \tag{9.2}$$

where $u \ll 1$ is the small parameter that describes the strength of disorder, and p_0 is the normalization constant.

Now, if one is interested in the critical properties of the system, it is necessary to integrate over all local field configurations up to the scale of the correlation length. This type of calculation is usually performed using a renormalization-group (RG) scheme, which self-consistently takes into account all the fluctuations of the field on scale lengths up to R_c.

In order to derive the traditional results for the critical properties of this system one can use the usual RG procedure developed for dimensions $D = 4 - \epsilon$, where $\epsilon \ll 1$. Then one finds that in the presence of the quenched disorder the pure system fixed point becomes unstable, and the RG rescaling trajectories arrive to another (universal) fixed point $g_* \neq 0$; $u_* \neq 0$, which yields the new critical exponents describing the critical properties of the system with disorder.

However, there exists an important point, which is missing in the traditional approach. Consider the ground-state properties of the system described by the Hamiltonian (9.1). Configurations of the fields $\phi(x)$ which correspond to local minima in H satisfy the saddle-point equation:

$$-\Delta\phi(x) + (\tau - \delta\tau(x))\phi(x) + g\phi^3(x) = 0 \tag{9.3}$$

Clearly, the solutions of this equation depend on a particular configuration of the function $\delta\tau(x)$ being inhomogeneous. The localized solutions with non-zero values of ϕ exist in regions of space where $(\tau - \delta\tau(x))$ has negative values. Moreover, one finds a *macroscopic* number of local minimum solutions of the saddle-point equation (9.3). Indeed, for a given realization of the random function $\delta\tau(x)$ there exists a macroscopic number of spatial 'islands' where $(\tau - \delta\tau(x))$ is negative (so that the local effective temperature is below T_c), and in each of these 'islands' one finds two local minimum configurations of the field: one which is 'up', and another which is 'down'. These local minimum energy configurations are separated by finite energy barriers, whose heights increase as the sizes of the 'islands' are increased.

The problem is that the traditional RG approach is only a perturbative theory in which the deviations of the field around the ground-state configuration are treated, and it can not take into account other local minimum configurations which are 'beyond barriers'. This problem does not arise in the pure systems, where the solution of the saddle-point equation is unique. However, in a situation such as that discussed above, when one gets numerous local minimum configurations separated by finite barriers, the direct application of the traditional RG scheme may be questioned.

In a systematic approach one would like to integrate in an RG way over fluctuations around the local minima configurations. Furthermore, one also has to sum over all these local minima up to the scale of the correlation length. In view of the fact that the local minima configurations are defined by the random quenched function $\delta\tau(x)$ in an essentially non-local way, the possibility of implementing such a systematic approach successfully seems rather hopeless.

On the other hand there exists another technique, which has been developed specifically for dealing with systems that exhibit numerous local minima states. It is the Parisi replica symmetry breaking (RSB) scheme, which has proved to be crucial in the mean-field theory of spin glasses (see Chapters 3–5). Recent studies show that in certain cases the RSB approach can also be generalized for situations where one has to deal with fluctuations as well [36–38]. Moreover, recently it has been shown that the RSB technique can be applied successfully for the RG studies of the critical phenomena in the Sine–Gordon model where remarkable instability of the RG flows with respect to the RSB modes has been discovered [39]. It can be argued that the summation over multiple local minimum configurations in the present problem could provide additional non-trivial RSB interaction potentials for the fluctuating fields [40]. Let us consider this point in more detail.

To carry out the appropriate average over quenched disorder one can use the standard replica approach (Sections 1.3 and 8.2). This is accomplished by introducing the replicated partition function, $Z_n \equiv \overline{Z^n[\delta\tau]}$, which is defined by the following replica Hamiltonian:

$$H_n = -\int d^D x \left[\frac{1}{2}\sum_{a=1}^{n}(\nabla\phi_a(x))^2 + \frac{1}{2}\tau\sum_{a=1}^{n}\phi_a^2(x) + \frac{1}{4}\sum_{a,b=1}^{n} g_{ab}\phi_a^2(x)\phi_b^2(x) \right] \tag{9.4}$$

Here

$$g_{ab} = g\delta_{ab} - u \tag{9.5}$$

is the *replica symmetric* (RS) coupling matrix. If one starts the usual RG procedure for the above replica Hamiltonian (as in Section 8.2), then it would correspond to the perturbation theory around the homogeneous ground state $\phi = 0$.

However, in the situation when there exist numerous local minima solutions of the saddle-point equation (9.3) we have to be more careful. Let us denote the local solutions of Eq. (9.3) by $\psi^{(i)}(x)$ where $i = 1, 2, \ldots, N_0$ labels the 'islands' where $\delta\tau(x) > \tau$. If the size L_0 of an 'island' where $(\delta\tau(x) - \tau) > 0$ is not too small, then the value of $\psi^{(i)}(x)$ in this 'island' should be $\sim \pm\sqrt{(\delta\tau(x) - \tau)/g}$, where $\delta\tau(x)$ should now be interpreted as the value of $\delta\tau$ averaged over the region of size L_0. Such 'islands' occur at a certain finite density per unit volume. Thus the value of N_0 is macroscopic: $N_0 = \kappa V$, where V is the volume of the system and κ is a constant. An approximate global extremal solution $\Phi(x)$ is constructed as the union of all these local solutions, and each local solution can occur with either sign:

$$\Phi_{(\alpha)}[x; \delta\tau(x)] = \sum_{i=1}^{\kappa V} \sigma_i \psi^{(i)}(x) \tag{9.6}$$

where each $\sigma_i = \pm 1$. Accordingly, the total number of global solutions must be $2^{\kappa V}$. We label these solutions by $\alpha = 1, 2, \ldots, K = 2^{\kappa V}$. As mentioned earlier, it seems unlikely that an integration over fluctuations around $\phi(x) = 0$ will include the contributions from the configurations of $\phi(x)$ that are near $\Phi(x)$, because $\Phi(x)$ is 'beyond a barrier', so to speak. Therefore, it seems appropriate to include separately the contributions from small fluctuations about each of the many $\Phi_{(\alpha)}[x; \delta\tau]$. Thus we have to sum over the K global minimum solutions (non-perturbative degrees of freedom) $\Phi_{(\alpha)}[x; \delta\tau]$ and also to integrate over 'smooth' fluctuations $\varphi(x)$ around them

$$\begin{aligned} Z[\delta\tau] &= \int D\varphi(x) \sum_{\alpha}^{K} \exp\left(-H[\Phi_{(\alpha)} + \varphi; \delta\tau]\right) \\ &= \int D\varphi(x) \exp\left(-H[\varphi; \delta\tau]\right) \times \tilde{Z}[\varphi; \delta\tau] \end{aligned} \tag{9.7}$$

where

$$\tilde{Z}[\varphi; \delta\tau] = \sum_{\alpha}^{K} \exp\left(-H_\alpha - \int d^D x \left[\frac{3}{2} g \Phi^2_{(\alpha)}(x; \delta\tau)\varphi^2(x) + g\Phi_{(\alpha)}(x; \delta\tau)\varphi^3(x)\right]\right) \tag{9.8}$$

and H_α is the energy of the α-th solution. Next we carry out the appropriate average over quenched disorder, and for the replica partition function, Z_n, we get:

$$Z_n = \int D\delta\tau P[\delta\tau] \int D\varphi_a \exp\left(-\sum_{a=1}^{n} H[\varphi_a; \delta\tau]\right) \times \tilde{Z}_n[\varphi_a; \delta\tau] \tag{9.9}$$

where the subscript a is a replica index and

$$\begin{aligned} \tilde{Z}_n[\varphi_a; \delta\tau] = \sum_{\alpha_1 \ldots \alpha_n}^{K} \exp\Bigg(&-\sum_{a}^{n} H_{\alpha_a} \\ &- \int d^D x \sum_{a}^{n} \left[\frac{3}{2} g \Phi^2_{(\alpha_a)}(x; \delta\tau)\varphi_a^2(x) + g\Phi_{(\alpha_a)}(x; \delta\tau)\varphi_a^3(x)\right]\Bigg) \end{aligned} \tag{9.10}$$

It is clear that if the saddle-point solution is unique, from Eqs. (9.9) and (9.10) one would obtain the usual RS representation (9.4) and (9.5). However, in the case of the macroscopic number of local minimum solutions the problem becomes highly non-trivial.

It is obviously hopeless to try to make a systematic evaluation of the above replicated partition function. The global solutions $\Phi^{(\alpha)}$ are complicated implicit functions of $\delta\tau(x)$. These quantities have fluctuations of two different types. In the first instance, they depend on the stochastic variable $\delta\tau(x)$. But even when the $\delta\tau(x)$ is fixed, $\Phi_{(\alpha)}(x)$ will depend on α (which labels the possible ways of constructing the global minimum out of the choices for the signs $\{\sigma\}$ of the local minima). A crude way of treating this situation is to regard the local solutions $\psi^{(i)}(x)$ as if they were random variables, even though $\delta\tau(x)$ has been specified. This randomness, which one can see is not all that different from that which exists in spin glasses, is the crucial one. It can be argued then, that owing to the interaction of the fluctuating fields with the local minima configurations (the term $\Phi^2_{(\alpha_a)}\varphi^2_a$ in Eq. (9.10)), the summation over solutions in the replica partition function $\tilde{Z}_n[\varphi_a]$, (Eq. (9.10)), in principle could provide the additional non-trivial RSB potential

$$\sum_{a,b} g_{ab}\varphi^2_a\varphi^2_b$$

in which the matrix g_{ab} has the Parisi RSB structure [40].

One may look at this problem another way. In terms of the direct replica approach the system under consideration is described by the replica Hamiltonian, Eq. (9.4). This is the *exact* result of the averaging over random $\delta\tau(x)$, and the coupling matrix g_{ab}, (Eq. (9.5)), in this Hamiltonian contains no replica symmetry breaking. Nevertheless, although in this representation the system is homogeneous and it contains now no spatial 'islands' discussed above, the problem of numerous non-trivial localized solutions exists here too. It can be shown (see Chapter 13) that besides the trivial configuration $\phi_a = 0$, the corresponding saddle-point equation:

$$-\Delta\phi_a(x) + \tau\phi_a(x) + g\phi^3_a(x) - u\phi_a(x)\left(\sum_{b=1}^{n}\phi^2_b(x)\right) = 0 \qquad (9.11)$$

has an infinite number of non-trivial instanton-like solutions having finite spatial size and finite energy. These solutions are characterized by explicit replica symmetry breaking in the replica *vector* ϕ_a, and therefore after proper summation over all such solutions the resulting effective interactions of the remaining fluctuating fields could, in principle, contain some kind of the replica symmetry breaking structure. Unfortunately, a systematic way of summation over all these solutions has not yet been worked out.

In this chapter we are going to study the critical properties of weakly disordered systems in terms of the RG approach, taking into account the possibility of a general type of RSB potential for fluctuating fields. The idea is that hopefully, as in spin glasses, this type of generalized RG scheme self-consistently takes into account relevant degrees of freedom coming from the numerous local minima. In particular, the instability of the traditional RS fixed points with respect to the RSB indicates that the multiplicity of the local minima can be relevant for the critical properties in the fluctuation region.

It will be shown (in Section 9.2) that, whenever the disorder is relevant for the critical behavior, the usual RS fixed points (which used to be considered as providing new universal disorder-induced critical exponents) are unstable with respect to 'turning on' an RSB potential. Moreover, it will be shown that in the presence of a general type of RSB potential, the RG flows actually lead to the so-called *strong coupling regime* at the finite spatial scale $R_* \sim \exp(1/u)$ (which corresponds to the temperature scale $\tau_* \sim \exp(-1/u)$). At this scale the renormalized matrix g_{ab} develops strong RSB, and the values of the interaction parameters are no longer small [41].

Usually the strong coupling situation indicates that certain essentially non-perturbative excitations have to be taken into account, and it could be argued that in the present model these are caused by exponentially rare instanton-like solutions of the saddle-point equation (9.11). A distant analog of this situation exists in the two-dimensional Heisenberg model where the Polyakov renormalization develops into the strong coupling regime at a finite (exponentially large) scale, which is known to be caused by the non-linear localized instanton solutions [42].

9.2 Replica symmetry breaking in the RG theory

Let us again consider the p-component ferromagnet with quenched random effective temperature fluctuations described by the usual Ginzburg–Landau Hamiltonian, Eq. (8.14). In terms of the standard replica approach after integration over the disorder variable $\delta\tau(x)$ for the corresponding replica Hamiltonian we get:

$$H_n = \int d^D x \left[\frac{1}{2} \sum_{i=1}^{p} \sum_{a=1}^{n} (\nabla \phi_i^a(x))^2 + \frac{1}{2} \tau \sum_{i=1}^{p} \sum_{a=1}^{n} (\phi_i^a(x))^2 + \frac{1}{4} \sum_{i,j=1}^{p} \sum_{a,b=1}^{n} g_{ab} (\phi_i^a(x))^2 (\phi_j^b(x))^2 \right] \quad (9.12)$$

Along the lines of the usual rescaling scheme at dimension $D = 4-\epsilon$ (Section 8.2) one gets the following (one-loop) RG equations for the interaction parameters g_{ab} (see Eq. (8.20)):

$$\frac{dg_{ab}}{d\xi} = \epsilon g_{ab} - \frac{1}{8\pi^2}\left[4g_{ab}^2 + 2(g_{aa}+g_{bb})g_{ab} + p\sum_{c=1}^{n} g_{ac}g_{cb}\right] \tag{9.13}$$

Redefining $g_{ab} \to 8\pi^2 g_{ab}$, and $g_{a\neq b} \to -g_{a\neq b}$ (so that the off-diagonal elements would be positively defined), and introducing the notation $\tilde{g} \equiv g_{aa}$, we get the RG equations in the following form:

$$\frac{dg_{ab}}{d\xi} = \epsilon g_{ab} - (4+2p)\tilde{g}g_{ab} + 4g_{ab}^2 + p\sum_{c\neq a,b}^{n} g_{ac}g_{cb} \qquad (a\neq b) \tag{9.14}$$

$$\frac{d}{d\xi}\tilde{g} = \epsilon\tilde{g} - (8+p)\tilde{g}^2 - p\sum_{c\neq 1}^{n} g_{1c}^2 \tag{9.15}$$

Using these equations for the replica-symmetric matrix g_{ab} one can easily recover the usual RG equations (8.22) of Section 8.2. Here we shall study Eqs. (9.14) and (9.15) assuming that the matrix g_{ab} has a general Parisi RSB structure.

According to the standard technique of the Parisi RSB algebra (see Section 3.4), in the limit $n \to 0$ the matrix g_{ab} is parametrized in terms of its diagonal elements $\tilde{g}$ and the off-diagonal *function* $g(x)$ defined in the interval $0 < x < 1$. All the operations with such matrices can be performed according to the following simple rules (see Eqs. (3.40)–(3.43)):

$$g_{ab}^k \to (\tilde{g}^k; g^k(x)) \tag{9.16}$$

$$(\hat{g}^2)_{ab} \equiv \sum_{c=1}^{n} g_{ac}g_{cb} \to (\tilde{c}; c(x)) \tag{9.17}$$

where

$$\tilde{c} = \tilde{g}^2 - \int_0^1 dx g^2(x)$$

$$c(x) = 2\left(\tilde{g} - \int_0^1 dy g(y)\right) g(x) - \int_0^x dy\,[g(x) - g(y)]^2 \tag{9.18}$$

The RS situation corresponds to the case $g(x) = \text{const}$, independent of x.

Using the above rules, from Eqs. (9.14) and (9.15) one gets:

$$\frac{d}{d\xi}g(x) = [\epsilon - (4+2p)\tilde{g}]\,g(x) + 4g^2(x)$$

$$-2pg(x)\int_0^1 dyg(y) - p\int_0^x dy\,[g(x)-g(y)]^2 \qquad (9.19)$$

$$\frac{d}{d\xi}\tilde{g} = \epsilon\tilde{g} - (8+p)\tilde{g}^2 + p\overline{g^2} \qquad (9.20)$$

where $\overline{g^2} \equiv \int_0^1 dxg^2(x)$.

Usually in the studies of the critical behavior one is looking for the stable fixed-point solution of the RG equations. From Eq. (9.19) one can easily determine the structure of the function $g(x)$ in the fixed point that is defined by the equations $\frac{d}{d\xi}g(x) = 0$, $\frac{d}{d\xi}\tilde{g} = 0$. Taking the derivative over x twice, from the r.h.s. of Eq. (9.19) one easily finds that in the fixed point $\frac{d}{dx}g(x) = 0$. This means that either the function $g(x)$ is a constant (which is the RS situation), or it has the step-like structure. It is interesting to note that the structure of this fixed-point equation is similar to that for the Parisi function $q(x)$ near T_c in the Potts spin glasses [43], and it is the term $g^2(x)$ in Eq. (9.19) that is known to produce one-step RSB solution there. The numerical solution of the RG equations given above demonstrates convincingly that whenever the trial function $g(x)$ has the many-step RSB structure, it quickly develops into the one-step one with the coordinate of the step being the most-right step of the original many-step function.

Let us consider the one-step RSB ansatz for the fixed-point function $g(x)$:

$$g(x) = \begin{cases} g_0 & \text{for } 0 \le x < x_0 \\ g_1 & \text{for } x_0 < x \le 1 \end{cases} \qquad (9.21)$$

where $0 \le x_0 \le 1$ is a (free) parameter that characterizes the structure of a particular fixed-point function. In terms of this ansatz from Eqs. (9.19) and (9.20) one easily gets the following fixed-point equations for the parameters g_1, g_0 and $\tilde{g}$:

$$\begin{aligned} &(4-2px_0)g_0^2 - 2p(1-x_0)g_1g_0 - (4+2p)\tilde{g}g_0 + \epsilon g_0 = 0 \\ &-px_0g_0^2 + (4-2p+px_0)g_1^2 - (4+2p)\tilde{g}g_1 + \epsilon g_1 = 0 \\ &-px_0g_0^2 - p(1-x_0)g_1^2 + (8+p)\tilde{g}^2 - \epsilon\tilde{g} = 0 \end{aligned} \qquad (9.22)$$

It should be stressed that unlike the corresponding saddle-point equations in spin glasses, here we have *no condition* for the parameter x_0. Let us recall, that the saddle-point equations appear as the result of optimization of an energy function with respect to *all* parameters of the considered ansatz solution. On the other hand, here we have the 'dynamical' evolution equations, and the parameters of a particular solution, in principle, can be defined by the initial conditions. Of course, in the search for a fixed-point solution one can hope to find a 'universal' (independent of the initial conditions) result, which indeed takes place in the RS situation considered in the previous chapter. Here, instead, we face the situation that, in the framework of the considered 'dynamics', one of the parameters of the ansatz function *is not evolving*, being defined *only* by the initial conditions. If such a type of the stable fixed-point solution does exist (which indeed takes place, as we shall see below), generally speaking, it signals nothing less than the *breaking of universality*.

The fixed-point equations (9.22) have several non-trivial solutions:

(1) The RS fixed-point, which corresponds to the pure system, Eq. (8.27):

$$g_0 = g_1 = 0; \quad \tilde{g} = \frac{1}{8+p}\epsilon \tag{9.23}$$

One can check that this fixed point (in accordance with the Harris criterion) is stable for the number of spin components $p > 4$, and it becomes unstable for $p < 4$.

(2) The 'random' RS fixed point, Eq. (8.28), (for $p > 1$):

$$g_0 = g_1 = \epsilon\frac{4-p}{16(p-1)}; \quad \tilde{g} = \epsilon\frac{p}{16(p-1)} \tag{9.24}$$

This fixed point is usually considered to be the one which describes the new universal critical behavior in systems with disorder. In the framework of simple linear stability analysis it can be shown, however, that this fixed point is stable (for $1 < p < 4$) only with respect to the RS ($g_0 = g_1$) deviations. It can be relatively easily demonstrated that within the one-step RSB subspace the fixed point (9.24) is *always unstable* [40]. The three eigenvalues of the corresponding linearized equations near this fixed point are:

$$\lambda_1 = -1/2, \quad \lambda_2 = -\frac{(4-p)}{8(p-1)}, \quad \lambda_3 = +\frac{(4-p)}{8(p-1)}$$

so that one of these eigenvalues is always positive.

(3) The one-step RSB fixed point [40]:

$$\begin{aligned} g_0 &= 0 \\ g_1 &= \epsilon\frac{4-p}{16(p-1) - px_0(8+p)} \\ \tilde{g} &= \epsilon\frac{p(1-x_0)}{16(p-1) - px_0(8+p)} \end{aligned} \tag{9.25}$$

Slightly painful algebra of the linear stability analysis shows that this fixed point is stable (within the one-step RSB subspace!) in the following region of parameters:

$$1 < p < 4$$
$$0 < x_0 < x_c(p) \equiv \frac{16(p-1)}{p(8+p)} \qquad (9.26)$$

In particular, $x_c(p = 2) = 4/5$; $x_c(p = 3) = 32/33$, and $x_c(p = 4) = 1$. Using the result given by Eq. (9.25) one can easily obtain the corresponding critical exponents which become non-universal as they are dependent on the parameter x_0 (see Section 9.3). (Note that in addition to the fixed points listed above there exist several other one-step RSB solutions which are either unstable or unphysical.)

The problem, however, is that if the parameter x_0 of the starting function $g(x;\xi = 0)$ (or, more generally, the coordinate of the most-right step of the many-steps starting function) is taken to be beyond the stability interval, such that $x_c(p) < x_0 < 1$, then there exist *no stable fixed points* of the RG equations (9.19) and (9.20). One faces the same situation also in the case of a general continuous starting function $g(x;\xi = 0)$. Moreover, according to Eq. (9.26) there exist no stable fixed points (beyond the RS subspace) in the most interesting Ising case, $p = 1$. Unlike the RS situation, where for $p = 1$ one finds the stable $\sim\sqrt{\epsilon}$ fixed point in the two-loop RG equations, here adding next-order terms in the RG equations does not cure the problem. In the RSB case one finds that in the two-loop RG equations the values of the parameters in the fixed point are of the order of one, and this indicates that we are entering the strong coupling regime where all the orders of the RG becomes relevant.

Nevertheless, even in the absence of stable fixed points, to get indications at least about the tendencies of the critical behavior, one can still try to analyze the actual 'evolution' properties of the above one-loop RG equations. Until we reach a critical scale ξ_*, at which the strong coupling regime begins, the scale evolution of the parameters of the renormalized Hamiltonian still correctly describe the properties of the system.

The evolution of the renormalized function $g(x;\xi)$ can be analyzed both numerically and analytically. It can be shown (see [41]) that in the case $p < 4$ for a general continuous starting function $g(x;\xi = 0) \equiv g_0(x)$ the renormalized function $g(x;\xi)$ quickly tends to zero everywhere in the interval $0 \leq x < (1 - \Delta(\xi))$, whereas in the narrow (scale dependent) interval $\Delta(\xi)$

near $x = 1$ the typical value of the function $g(x;\xi)$ is divergent:

$$g(x;\xi) \sim \begin{cases} \dfrac{u}{1-u\xi} & \text{for } (1-x) \ll \Delta(\xi) \\ 0; & \text{for } (1-x) \gg \Delta(\xi) \end{cases} \tag{9.27}$$

$$\tilde{g}(\xi) \sim u \ln\frac{1}{1-u\xi} \tag{9.28}$$

where

$$\Delta(\xi) \simeq (1-u\xi) \tag{9.29}$$

According to this result the critical scale ξ_* is defined by the condition that the values of the renormalized parameters become of the order of one: $(1-u\xi_*) \sim u$, or $\xi_* \sim 1/u$. Correspondingly, the spatial scale at which the system enters the strong coupling regime is:

$$R_* \sim \exp\left(\frac{\text{const}}{u}\right) \tag{9.30}$$

Note that the value of this scale is much greater than the usual crossover scale $\sim u^{-\alpha/\nu}$ (where α and ν are the pure system specific heat and the correlation length critical exponents), at which the disorder becomes relevant for the critical behavior.

Qualitatively similar asymptotic behavior of the function $g(x;\xi)$ is obtained for the case when the starting function $g_0(x)$ has the one-step RSB structure (9.21) with the coordinate of the step x_0 in the 'instability region' $(16(p-1)/p(8+p) \equiv x_c(p) < x_0 < 1)$:

$$g(x;\xi) \sim \begin{cases} \dfrac{g_1(0)}{1-(4-2p+px_0)g_1(0)\xi}; & \text{for } x_0 < x < 1 \\ 0; & \text{for } 0 \leq x < x_0 \end{cases} \tag{9.31}$$

Here $g_1(0) \equiv g_1(\xi = 0) \sim u$, and the coefficient $(4-2p+px_0)$ is always positive. In this case the system also enters into the strong coupling regime at scales $\xi \sim 1/u$.

Note that the above asymptotics do not explicitly involve ϵ. In fact the role of the parameter $\epsilon > 0$ is to 'push' the RG trajectories out of the trivial Gaussian fixed point $g = \tilde{g} = 0$. Thus, the value of ϵ, as well as the values of the starting parameters $g_0(x)$, $\tilde{g}_0$, define a scale at which the solutions finally enter the above asymptotic regime. When $\epsilon < 0$ (above dimension 4), the Gaussian fixed point is stable; on the other hand, the strong coupling asymptotics still exists in this case as well, separated from the trivial one by a finite (depending on the value of ϵ) barrier. Therefore, although *infinitely small* disorder remains irrelevant for the critical behavior above the dimension 4, if the disorder is strong enough (bigger than some value depending on the ϵ threshold) the RG trajectories could enter the strong coupling regime too.

9.3 Scaling properties and the replica symmetry breaking

9.3.1 Spatial and temperature scales

The renormalization of the mass term

$$\tau(\xi)\sum_{a=1}^{n}\phi_a^2$$

is described by the following RG equation (see Eq. (8.24)):

$$\frac{d}{d\xi}\ln|\tau| = 2 - \frac{1}{8\pi^2}\left[(2+p)\tilde{g} + p\sum_{a\neq 1}^{n} g_{1a}\right] \tag{9.32}$$

After the redefinition (as in the previous section) $g_{ab} \to 8\pi^2 g_{ab}$, and $g_{a\neq b} \to -g_{a\neq b}$, in the Parisi representation we get:

$$\frac{d}{d\xi}\ln|\tau| = 2 - [(2+p)\tilde{g}(\xi) + p\int_0^1 g(x;\xi)] \tag{9.33}$$

or

$$\tau(\xi) = \tau_0 \exp\left(2\xi - \int_0^{\xi} d\eta[(2+p)\tilde{g}(\eta) + p\overline{g}(\eta)]\right) \tag{9.34}$$

where $\tilde{g}(\eta)$ and $\overline{g}(\eta) \equiv \int_0^1 dx g(x;\eta)$ are the solutions of the RG equations of the previous section.

Let us first consider the traditional (replica-symmetric) situation. The RS interaction parameters $\tilde{g}(\xi)$ and $g(\xi)$ approach the fixed point values $\tilde{g}_*$ and g_* (which are of the order of ϵ), and then for the dependence of the renormalized mass $\tau(\xi)$, according to (9.34), one gets:

$$\tau(\xi) = \tau_0 \exp\{\Delta_\tau \xi\} \tag{9.35}$$

where

$$\Delta_\tau = 2 - [(2+p)\tilde{g}_* + pg_*] \tag{9.36}$$

At scale ξ_c, such that $\tau(\xi_c)$ becomes of the order of one, the system gets out of the scaling region, and this defines the correlation length R_c as a function of the reduced temperature τ_0. Because $\xi = \ln R$, according to (9.35), we obtain:

$$R_c(\tau_0) \sim \tau_0^{-\nu} \tag{9.37}$$

where $\nu = 1/\Delta_\tau$ is the critical exponent of the correlation length.

Actually, if the starting value of the disorder parameter $g(\xi = 0) \equiv u$ is much smaller than the starting value of the pure system interaction $\tilde{g}(\xi = 0) \equiv g_0$, the situation is a little bit more complicated. In this case the RG flow for $\tilde{g}(\xi)$ first arrives at the pure system fixed point $\tilde{g}_*^{(\text{pure})}$, as if the disorder perturbation does not exist. Then, because the pure fixed point is unstable with respect to the disorder perturbations, at scales bigger than certain disorder dependent scale ξ_u the RG trajectories eventually arrive at the stable (universal) 'random' fixed point $(\tilde{g}_*, g_*)$. According to the traditional theory [35] it is known that $\xi_u \sim (\nu/\alpha)\ln(1/u)$, and the corresponding spatial scale is $R_u \sim u^{-\nu/\alpha}$.

Coming back to the scaling behavior of the mass parameter $\tau(\xi)$, (Eq. (9.35)), we see that if the value of the temperature τ_0 is such that $\tau(\xi)$ becomes of the order of one before the crossover scale ξ_u is reached, then for the scaling behavior of the correlation length (as well as for other thermodynamic quantities) one finds for the pure system the result $R_c(\tau_0) \sim \tau_0^{-\nu(\text{pure})}$. However, critical behavior of the pure system is observed only until $R_c \ll R_u$, which imposes the following restriction on the temperature parameter: $\tau_0 \gg u^{1/\alpha} \equiv \tau_u$. In other words, at temperatures not too close to T_c, $\tau_u \ll \tau_0 \ll 1$, the presence of disorder is irrelevant for the critical behavior.

On the other hand, if $\tau_0 \ll \tau_u$ (in the close vicinity of T_c), the RG trajectories for $\tilde{g}(\xi)$ and $g(\xi)$ arrive (after crossover) at a new (universal) 'random' fixed point $(\tilde{g}_*, g_*)$, and the scaling of the correlation length (as well as other thermodynamic quantities), according to Eqs. (9.36)–(9.37), is controlled by a new universal critical exponent ν, which is defined by the RS fixed point $(\tilde{g}_*, g_*)$ of the random system.

Consider now the situation in the RSB case. Again, if the disorder parameter u is small, in the temperature interval $\tau_u \ll \tau_0 \ll 1$, the critical behavior is essentially controlled by the 'pure' fixed point, and the presence of disorder is irrelevant. For the same reasons as discussed above, the system gets out of the scaling regime ($\tau(\xi)$ becomes of the order of one) before the disorder parameters start 'pushing' the RG trajectories out of the pure system fixed point.

However, at temperatures $\tau_0 \ll \tau_u$ the situation is completely different from the RS case. If the RG trajectories arrive at the one-step RSB fixed point, (Eq. (9.25)), then (in the $1 < p < 4$ case) according to the standard scaling relations for the critical exponent of the correlation length one finds:

$$\nu(x_0) = \frac{1}{2} + \frac{1}{2}\epsilon \frac{3p(1 - x_0)}{16(p-1) - px_0(p+8)} \tag{9.38}$$

Thus, depending on the value of the parameter x_0 one finds a whole *spectrum* of the critical exponents. Therefore, unlike the traditional point of view described in Section 8.2, the critical properties become *non-universal*, because they are dependent on the concrete statistical properties of the disorder involved (which eventually define the value of x_0). However, this result is not the only consequence of the RSB. More essential effects can be observed in the scaling properties of the spatial correlation functions (see below).

In the Ising case, $p = 1$, as well as for a generic continuous starting function $g_0(x)$, the consequences of the RSB appear to be even more dramatic. Here, at scales $\xi \gg \xi_u$ (although still $\xi \ll \xi_* \sim 1/u$) according to the solutions (9.27) and (9.31) the parameters $\tilde{g}(\xi)$ and $g(x;\xi)$ do not arrive at any fixed point, and they keep evolving as the scale ξ increases. Therefore, in this case, according to Eq. (9.34), the correlation length (defined, as usual, by the condition that the renormalized $\tau(\xi)$ becomes of the order of one) is defined by the following non-trivial equation:

$$2\ln R_c - \int_0^{\ln R_c} d\eta \, [(2+p)\tilde{g}(\eta) + p\overline{g}(\eta)] = \ln\frac{1}{\tau_0} \tag{9.39}$$

Thus, as the temperature becomes sufficiently close to T_c (in the disorder dominated region $\tau_0 \ll \tau_u$) there will be *no usual scaling dependence* of the correlation length (as well as of other thermodynamic quantities).

Finally, if the temperature parameter τ_0 becomes sufficiently small, at the scale $\xi_* \equiv \ln R_* \sim 1/u$ the system enters the strong coupling regime (such that the parameters $\tilde{g}(\xi)$ and $g(x;\xi)$ are no longer small), while the renormalized mass $\tau(\xi)$ remains still small. The corresponding crossover temperature scale is:

$$\tau_* \sim \exp\left(-\frac{\text{const}}{u}\right) \tag{9.40}$$

In the close vicinity of T_c at $\tau \ll \tau_*$ we are facing the situation that at large scales the interaction parameters of the asymptotic (zero-mass) Hamiltonian are no longer small, and the properties of the system cannot be analysed in terms of simple RG analysis. Because it is the parameter describing the disorder, $g(x;\xi)$, that is the most divergent, in a sense the problem here is qualitatively reduced back to the original one but with *strong* disorder. These qualitative arguments indicate that in the narrow temperature interval $\tau \ll \tau_*$ near T_c at least some of the properties of the system may appear to be spin-glass-like.

9.3.2 Correlation functions

Consider the scaling properties of the spin-glass-type connected correlation function:

$$K(R) = \overline{[\langle\phi(0)\phi(R)\rangle - \langle\phi(0)\rangle\langle\phi(R)\rangle]^2} \equiv \overline{\langle\langle\phi(0)\phi(R)\rangle\rangle^2} \tag{9.41}$$

In terms of the replica formalism it can be represented as follows:

$$K(R) = \lim_{n\to 0} \frac{1}{n(n-1)} \sum_{a\neq b}^{n} K_{ab}(R) \tag{9.42}$$

where

$$K_{ab}(R) = \langle\langle\phi_a(0)\phi_b(0)\phi_a(R)\phi_b(R)\rangle\rangle \tag{9.43}$$

In terms of the standard RG formalism for the replica correlation function $K_{ab}(R)$ we obtain:

$$K_{ab}(R) \sim (G_0(R))^2(Z_{ab}(R))^2 \tag{9.44}$$

where

$$G_0(R) = R^{-(D-2)} \tag{9.45}$$

is the free-field correlation function, and in the one-loop approximation the scaling of the mass-like object $Z_{ab}(R)$ (with $a \neq b$) is defined by the RG equation:

$$\frac{d}{d\xi}\ln(Z_{ab}(\xi)) = 2g_{ab}(\xi) \tag{9.46}$$

with initial condition $Z_{ab}(\xi = 0) = 1$. Here $g_{a\neq b}(\xi) > 0$ is the solution of the corresponding RG equations (9.14)–(9.15). According to Eqs. (9.44)–(9.46) we find:

$$K_{ab}(R) \sim R^{-2(D-2)} \exp\left[4\int_0^{\ln R} d\xi g_{ab}(\xi)\right] \tag{9.47}$$

Correspondingly, in the Parisi representation: $g_{a\neq b}(\xi) \to g(x;\xi)$ and $K_{a\neq b}(R) \to K(x;R)$, we get:

$$K(x;R) \sim R^{-2(D-2)} \exp\left[4\int_0^{\ln R} d\xi g(x;\xi)\right] \tag{9.48}$$

Let us first consider the traditional RS situation. In this case (for $1 < p < 4$) the interaction parameter $g_{a\neq b}(\xi) \equiv u(\xi)$ arrives to the RS fixed point

$$u_* = \epsilon\frac{4-p}{16(p-1)}$$

and according to Eqs. (9.47) and (9.42) we obtains the following simple scaling:

$$K_{rs}(R) \sim R^{-2(D-2)+\theta} \tag{9.49}$$

with the universal 'random' critical exponent

$$\theta = \epsilon \frac{4-p}{4(p-1)} \tag{9.50}$$

In the case of the one-step RSB fixed point, (Eq. (9.25)), the situation is somewhat more complicated. Here one finds that the correlation function $K(x;R)$ also has one-step RSB structure:

$$K(x;R) = \begin{cases} K_0(R); & \text{for } 0 \leq x < x_0 \\ K_1(R); & \text{for } x_0 < x \leq 1 \end{cases} \tag{9.51}$$

where (in the first order in ϵ)

$$\begin{aligned} K_0(R) &\sim R^{-2(D-2)} \equiv G_0^2(R) \\ K_1(R) &\sim R^{-2(D-2)+\theta_{1rsb}} \end{aligned} \tag{9.52}$$

with the *non-universal* critical exponent θ_{1rsb} explicitly depending on the parameter x_0:

$$\theta_{1rsb} = \epsilon \frac{4(4-p)}{16(p-1) - px_0(8+p)} \tag{9.53}$$

Because the critical exponent θ_{1rsb} is positive, the asymptotic scaling behavior of the 'observable' quantity $K(R) = \overline{\langle\langle\phi(0)\phi(R)\rangle\rangle^2}$, (Eq. (9.42)), is defined only by $K_1(R)$:

$$K(R) = \int_0^1 dx K(x;R) = (1-x_0)K_1(R) + x_0 K_0(R) \sim R^{-2(D-2)+\theta_{1rsb}} \tag{9.54}$$

It should be stressed, however, that the difference of the one-step RSB situation from the RS case manifests itself not only in the result that its critical exponents θ_{1rsb} is non-universal. According to the physical interpretation of the replica theory of spin glasses (Chapter 4), in the situation that a given replica quantity appears to be dependent on the replica indices, one has to expect that the observable value of the corresponding physical quantity must remain *probabilistic* even in the thermodynamic limit. Of course, in usual experiments one is measuring spatially averaged quantities. In particular, the 'observable' value of the two-point correlation functions (9.41) (in a given sample) is obtained by integration over the positions of the two points, such that the distance R between them is fixed, and this procedure is equivalent to the averaging over different realizations of disorder. In terms of the replica theory, the quantity obtained in this way is defined by Eq. (9.54), and of course it is not probabilistic.

Nevertheless, for a somewhat different scheme of the measurements, the qualitative difference from the RS situation can be observed. If we adopt for the present case the usual interpretation of the RSB phenomenon as factorization of the phase space into 'valleys' separated by big barriers of the free energy, then one could expect that besides the usual (slow) thermalization inside a particular valley, qualitatively much bigger relaxation times would be required for overcoming barriers separating different valleys. Therefore, the measurements in the 'thermal equilibrium' at not very long times can in fact correspond to the equilibration within one valley only and not to the true thermal equilibrium. Then in different measurements the system can be considered to be effectively 'trapped' in different valleys and thus the traditional spin-glass situation is recovered. Therefore, like in spin glasses, we can consider the 'overlap' quantities, which are the usual instrument for describing the RSB phenomena. For instance, instead of the correlation function (9.41) one can introduce the following spatially averaged quantity, which describes the overlap of the values of the spin–spin correlation function obtained in two different measurements in a given sample:

$$K_{ij}(R) \equiv \frac{1}{V} \int d^D r \langle \phi(r)\phi(r+R)\rangle_i \langle \phi(r)\phi(r+R)\rangle_j \qquad (9.55)$$

where i and j label different measurements, and it is assumed that the measurable thermal average corresponds to a particular valley, and not to the true thermal equilibrium. Now, if the RS situation occurs (so that the free energy has only one global valley), then the above quantity is equivalent to the usual correlation function (9.41), and for any pair of measurements one will obtain the same result as given by Eq. (9.49). On the other hand, in the case of the one-step RSB, the quantity (9.55) must be fluctuating from pair to pair of measurements. Moreover, taking into account the result (9.51), according to the general arguments of the RSB theory (Chapter 4), it can be predicted that the value $K_0(R)$ must be obtained with the probability x_0, while the value $K_1(R)$ must be obtained with the probability $(1 - x_0)$.

Consider finally the situation if a general type of the RSB takes place. According to the qualitative solution (9.27)–(9.28), the function $g(x;\xi)$ does not arrive at any fixed point. Therefore, according to Eq. (9.48), at the disorder dominated scales $R \gg R_u \sim u^{-\nu/\alpha} \gg 1$ there will be no simple scaling behavior of the correlation function $K(R)$. Finally, at scales $R \gg R_*$ the system enters into the strong coupling regime, where in the framework of the present RG approach no predictions for the behavior of the correlation functions can be made.

9.3.3 *Specific heat*

As usual the leading singularity of the specific heat can be calculated in terms of the following two-point correlation functions:

$$C \sim \int d^D R \left[\overline{\langle\phi^2(0)\phi^2(R)\rangle - \langle\phi^2(0)\rangle\langle\phi^2(R)\rangle} \right] \tag{9.56}$$

In terms of the RG scheme the correlation function:

$$W(R) \equiv \overline{\langle\phi^2(0)\phi^2(R)\rangle - \langle\phi^2(0)\rangle\langle\phi^2(R)\rangle} \tag{9.57}$$

can be represented as follows:

$$W(R) = G_0^2(R) m^2(R) \tag{9.58}$$

where $G_0(R) = R^{-(D-2)}$ is the free field two-point correlation function, and the scaling behavior of the mass-like object $m(R)$ is defined by the following (one-loop) RG equation (c.f. Eq. (9.33)):

$$\frac{d}{d\xi} \ln\left[m(\xi)\right] = -\left[(2+p)\tilde{g}(\xi) - p\sum_{a\neq 1}^{n} g_{a1}(\xi)\right] \tag{9.59}$$

Here, as usual, the renormalized parameters $\tilde{g}(\xi)$ and $g_{a\neq b}(\xi)$ are defined by the RG equations (9.14)–(9.15). In terms of the Parisi representation $g_{a\neq b}(\xi) \to g(x;\xi)$, the solution of the above equation is:

$$m(R) = \exp\left[-(2+p)\int_0^{\ln R} d\xi\, \tilde{g}(\xi) - p\int_0^{\ln R} d\xi \int_0^1 dx g(x;\xi)\right] \tag{9.60}$$

Substituting this result into Eqs. (9.58) and (9.56) after simple algebra for the singular part of the specific heat we get:

$$C \sim \int_0^{\xi_{\max}} d\xi \exp\left[\epsilon\xi - 2(2+p)\int_0^{\xi} d\eta\, \tilde{g}(\eta) - 2p\int_0^{\xi} d\eta \overline{g}(\eta)\right] \tag{9.61}$$

where $\overline{g}(\eta) \equiv \int_0^1 dx g(x;\eta)$, and the infrared cut-off $\xi_{\max}$ is the scale at which the system gets out of the scaling regime. In the usual scaling situation $\xi_{\max}$ is defined by the condition that the renormalized mass $\tau(\xi) \sim \tau_0 \exp(\Delta_\tau \xi)$, becomes of the order of one, so that $\xi_{\max} \sim \ln(1/\tau_0)$.

Again, let us first consider the RS case. Here at scales $\xi \gg \xi_u \sim \ln(1/u)$ (which correspond to the temperature region $\tau_0 \ll \tau_u \sim u^{1/\alpha}$) the renormalized parameters $\tilde{g}(\eta)$ and $g(\xi)$ arrive at the universal fixed point $\tilde{g}_* =$

$\epsilon[p/16(p-1)]$; $g_* = \epsilon(4-p)/[16(p-1)]$ (see Eq. (9.24)), and according to Eq. (9.61) for the singular part of the specific heat we get:

$$C(\tau_0) \sim \int_0^{\ln(1/\tau_0)} d\xi \exp\left[\xi\ (\epsilon - 2(2+p)\tilde{g}_* - 2pg_*)\right] \sim \tau_0^{\epsilon\frac{4-p}{8(p-1)}} \tag{9.62}$$

This singular behavior is described by the universal ('random') critical exponent $\alpha = -\epsilon(4-p)/[8(p-1)]$ (see Eq. (8.32)).

On the other hand, similar calculations for the one-step RSB fixed points, (Eq. (9.25)), yield the following *non-universal* specific heat critical exponent [40]:

$$\alpha(x_0) = -\frac{1}{2}\epsilon\frac{(4-p)(4-px_0)}{16(p-1)-px_0(p+8)}. \tag{9.63}$$

Thus, depending on the value of the parameter x_0 one finds a whole *spectrum* of the critical exponents:

$$-\infty < \alpha(x_0) < -\epsilon\frac{(4-p)}{8(p-1)}. \tag{9.64}$$

The upper bound for $\alpha(x_0)$ is achieved in the RS limit $x_0 \to 0$, and it coincides with the usual RS result, Eq. (8.32). On the other hand, as x_0 tends to the 'border of stability' $x_c(p)$ of the one-step RSB fixed point, formally the specific heat critical exponent tends to $-\infty$.

Finally, in the case of a general type of the RSB the situation becomes completely different. Here in the disorder dominated region $\tau_* \ll \tau_0 \ll u^{1/\alpha}$ (which corresponds to scales $\xi_u \ll \xi \ll \xi_*$) the RG trajectories of the interaction parameters $\tilde{g}(\xi)$ and $\overline{g}(\xi)$ do not arrive at any fixed point, and therefore, according to Eq. (9.62), one finds that the specific heat becomes a complicated function of the temperature parameter τ_0, which does not have the traditional scaling form. In the close vicinity of T_c, at $\tau_0 \ll \tau_*$, where the interaction parameters $\tilde{g}$ and $\overline{g}$ are becoming finite (formally, they are becoming divergent), one finds that the integral over ξ in Eq. (9.61) is becoming convergent (so that the upper temperature-dependent cutoff scale ξ_{max} becomes irrelevant). Thus, one arrives at the conclusion that the 'would be singular part' of the specific heat becomes *non-singular* in the narrow temperature interval $\tau_* \sim \exp\{-(\text{const})/u\}$ around T_c.

9.4 Discussion

According to the results obtained in this chapter, we can conclude that spontaneous replica symmetry breaking in the interactions of the critical fluctuations has a dramatic effect on the renormalization group flows and on the critical properties. In systems with the number of spin components $p < 4$ at dimension $D = 4 - \epsilon$ the usual replica-symmetric RG fixed points, which were believed to be describing the disorder-induced universal critical behavior, appear to be unstable with respect to tuning on the RSB potentials. For a general type of Parisi RSB structure there exist no stable fixed points, and the RG flows lead to the *strong coupling regime* at the finite scale $R_* \sim \exp(1/u)$ (which corresponds to the exponentially small temperature scale $\tau_* \sim \exp(-1/u)$), where u is the small parameter describing the disorder. Unlike the systems with $1 < p < 4$, where one finds stable fixed points having one-step RSB structures, (Eq. (9.25)), in the Ising case, $p = 1$, there exist no stable fixed points, and any RSB interactions lead to the strong coupling regime. All those effects lead either to the breaking of universality (in the case of one-step RSB fixed points), or to the breaking of scaling (in the case of the strong coupling regime).

The crucial general problem of this approach is that the convincing arguments demonstrating the origin of the spontaneous RSB in the interactions of critical fluctuations, are still lacking. To create firm ground for this type of theory one has to find a systematic way of summation over all non-trivial (instanton-like) solutions of the saddle-point equations. Hopefully this can be done in the framework of the vector replica symmetry breaking scheme, which will be described in Chapter 13.

The other key question that remains unanswered, is whether or not the obtained strong coupling phenomena in the RG theory could be interpreted as the onset of some kind of a 'glassy' phase in the narrow temperature interval $\tau_* \sim \exp(-1/u)$ around T_c. Because it is the coupling parameters describing the disorder that are the most divergent, it is tempting to argue that in this temperature interval the properties of the system should be spin-glass-like. It should be stressed, however, that in the present study we observe only the *crossover* temperature τ_*, at which the change of the critical regime occurs, and it is hardly possible to associate this temperature with any kind of thermodynamic phase transition. The true spin-glass order (in the traditional sense) arises from the onset of the non-zero *order parameter* $Q_{ab}(x) = < \phi_a(x)\phi_b(x) >; a \neq b$, and, at least for the infinite-range model, Q_{ab} develops the hierarchical dependence on replica indices (Chapter 3). In the present problem we only find that the *coupling matrix* g_{ab} of the fluctuating fields develops the RSB structure and its elements become non-small at large scales. Therefore, it seems more realistic to interpret the observed RSB strong coupling phenomena as some kind of a completely new critical behavior characterized by strong SG effects in the scaling properties of fluctuations rather than in the ground state.

There exists another general problem, which may appear to be interconnected with the RSB phenomena considered in this chapter. The problem is related to the existence of the so-called Griffiths phase [44] in a finite temperature interval above T_c. Numerous theoretical [45] and numerical [46] studies indicate that in the temperature interval $T_c < T < T_0$ (in the high-temperature phase) the time correlation functions decay as $\sim \exp\{-(t/\tau)^{\lambda}\}$ instead of the usual exponential relaxation law $\sim \exp\{-t/\tau\}$, as in the ordinary paramagnetic phase. It is claimed that the parameter $\lambda(T)$ is the continuous function of the temperature, such that $\lambda(T = T_c) < 1$, and it increases monotonically up to $\lambda(T = T_0) = 1$. The temperature T_0 coincides with the phase transition point of the corresponding pure system. Usually this phenomenon is associated with the existence of numerous local minima metastable states separated from the true ground state by big energy barriers.

On the other hand, according to the mean-field replica theory of spin glasses the RSB phenomena can be interpreted as a factorization of the phase space into a hierarchy of 'valleys' of local minima states, separated by macroscopic energy barriers. If such a general physical interpretation of the RSB effects can also be adopted for the systems considered here (with the reservation that the values of the corresponding energy barriers are finite and forming a continuous spectrum), then this qualitative physical picture would fit nicely with the Griffiths phenomena. Besides, the presence of the RSB in couplings of the fluctuating fields could be identified with a phase with a different symmetry than the conventional paramagnetic phase, and thus there would have to be a temperature T_{rsb} at which this change in symmetry occurs. On these grounds it is tempting to associate this (hypothetical) RSB transition (which must be the property of the statistics of the local minima saddle-point solutions) with the Griffiths transition at T_0.

10

Two-dimensional Ising model with disorder

10.1 Two-dimensional Ising systems

In the general theory of phase transitions, the two-dimensional (2D) Ising model plays the prominent role, as it is the simplest non-trivial lattice model with a known exact solution [47]. It is natural to ask, therefore, what the effect of quenched disorder is in this particular case. As for the Harris criterion (Section 8.1), the 2D Ising model constitutes a special case, because the specific heat exponent $\alpha = 0$ in this model. However, speaking intuitively, we could expect that as in the case of the vector field model in four dimensions (Section 8.3), the effect of disorder could be predicted on a qualitative level. Although the critical exponent α is zero, the specific heat of the 2D Ising model is (logarithmically) divergent at the critical point. Therefore, we should expect the critical behavior of this system to be strongly affected by the disorder.

Indeed, the exact solution for the critical behavior of the specific heat of the 2D Ising model with a small concentration $c \ll 1$ of impurities [48] (see Section 10.3 below) yields the following result for the singular part of the specific heat:

$$C(\tau) \sim \begin{cases} \ln\left(\dfrac{1}{\tau}\right) & \text{if } \tau^* \ll |\tau| \ll 1 \\ \ln\left[\ln\left(\dfrac{1}{\tau}\right)\right] & \text{if } \tau \ll \tau^* \end{cases} \tag{10.1}$$

where $\tau^* \sim \exp(-\text{const}/c)$ is the temperature scale at which a crossover from one critical behavior to another takes place.

Thus, in the 2D Ising model, as well as in the 4D vector field system, the disorder is relevant. However, unlike the vector field model, the specific heat of the 2D disordered Ising magnet remains divergent at T_c, though the singularity is weakened. Another important property of the 2D Ising model is that unlike the ϕ^4-theory near four dimensions (Chapter 9), the spin-glass RSB phenomena appear to be irrelevant for the critical behavior [52]. Thus, the result given by Eq. (10.1) for the leading singularity of the specific heat of the weakly disordered 2D Ising system must be exact.

In this chapter the emphasis is laid not on the exact lattice expressions, but on their large-scale asymptotics, i.e. we will be interested mainly in the critical long-range behavior because only that is interesting for the general theory of phase transitions. It is well known that in the critical region the 2D Ising model can be reduced to the free-fermion theory [11]. In Section 10.2 this reduction will be demonstrated in very simple terms by means of the Grassman variables technique. Neither the operator language nor the transfer matrix formalism will be used, as they are not symmetric enough to be applied to the model with disorder. The resulting continuum theory, to which the exact lattice disordered model is equivalent in the critical region, appears to be simple enough, and its specific heat critical behavior can be found exactly (Section 10.3).

The results of recent numerical simulations are briefly described in Section 10.4. The general structure of the phase diagram of the disordered 2D Ising model is considered in Section 10.5.

10.2 The fermion solution

The partition function of the pure 2D Ising model is given by:

$$Z = \sum_{\sigma} \exp\left[\beta \sum_{x,\mu} \sigma_x \sigma_{x+\mu}\right] \tag{10.2}$$

Here $\{\sigma_x = \pm 1\}$ are the Ising spins defined at lattice sites of a simple square lattice; x are integer valued coordinates of the lattice sites, and $\mu = \mathbf{1}, \mathbf{2}$ are basic vectors of the lattice.

This partition function can be rewritten as follows:

$$\begin{aligned} Z &= \sum_{\sigma} \prod_{x,\mu} \exp\left(\beta \sigma_x \sigma_{x+\mu}\right) = \sum_{\sigma} \prod_{x,\mu} \left(\cosh\beta + \sigma_x \sigma_{x+\mu} \sinh\beta\right) \\ &= (\cosh\beta)^V \sum_{\sigma} \prod_{x,\mu} \left(1 + \lambda \sigma_x \sigma_{x+\mu}\right) \end{aligned} \tag{10.3}$$

where V is the total number of the lattice bonds, and $\lambda \equiv \tanh\beta$. Expanding the product over the lattice bonds in Eq. (10.3) and averaging over the σs we obtain the following representation for the partition function (the

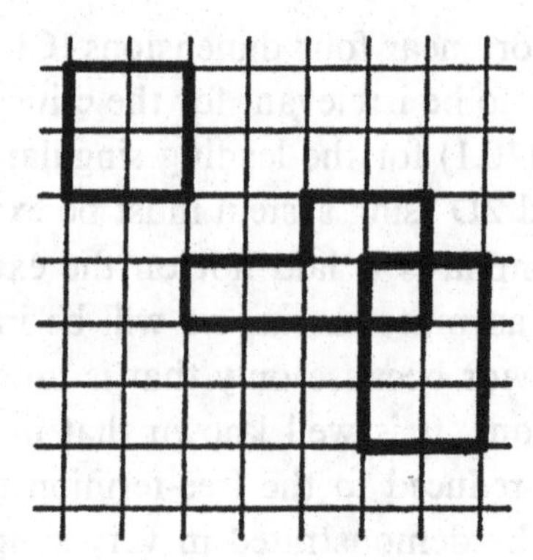

Fig. 10.1. Lattice graphs of the high-temperature expansion of the 2D Ising model.

high-temperature expansion):

$$Z = (\cosh\beta)^V \sum_{\mathscr{P}} (\lambda)^{L_{\mathscr{P}}} \tag{10.4}$$

The summation here goes over all configurations of closed paths $\mathscr{P}$ drawn on lattice links (Fig. 10.1), and $L_{\mathscr{P}}$ is the total length of paths in a particular configuration $\mathscr{P}$.

The summation in Eq. (10.4) can be performed exactly, and these calculations constitute the classical exact solution for the 2D Ising model found by Sherman and Vdovichenko [49]. This solution is described in detail in textbooks (see e.g. [50]), and we do not consider it here.

Let us now consider an alternative approach to the calculations of the partition function in terms of the so-called Grassmann variables (for detailed treatment of this new mathematics see [12]). The Grassmann variables were first used for the 2D Ising model by Hurst and Green [13], and this approach was later developed by a number of authors [14] (see also [48]). It appears that technically this method enables the equations to be obtained in a much simpler way. We shall describe this formalism, recover the equation for the partition function, Eq. (10.4), and introduce some new notations that will be useful for the problem with disorder.

Let us introduce the four-component Grassmann variables $\{\psi^\alpha(x)\}$ defined at the lattice sites $\{x\}$, where the superscript $\alpha = \mathbf{1, 2, 3, 4}$ indicates the four directions on the 2D square lattice (such that $\mathbf{3} \equiv \mathbf{-1}$ and $\mathbf{4} \equiv \mathbf{-2}$). All the $\{\psi^\alpha(x)\}$s and all their differentials $\{d\psi^\alpha(x)\}$ are anticommuting variables; by definition:

$$\psi^\alpha(x)\psi^\beta(y) = -\psi^\beta(y)\psi^\alpha(x)$$
$$(\psi^\alpha(x))^2 = 0$$

$$d\psi^\alpha(x)d\psi^\beta(y) = -d\psi^\beta(y)d\psi^\alpha(x)$$
$$d\psi^\alpha(x)\psi^\beta(y) = -\psi^\beta(y)d\psi^\alpha(x) \quad (10.5)$$

and the integration rules are defined as follows:

$$\int d\psi^\alpha(x) = 0$$
$$\int d\psi^\alpha(x)\psi^\alpha(x) = -\int \psi^\alpha(x)d\psi^\alpha(x) = 1 \quad (10.6)$$

Let us consider the following partition function defined as an integral over all the Grassmann variables of the 2D lattice system:

$$Z = \int D\psi \exp\{A[\psi]\} \quad (10.7)$$

Here the integration measure $D\psi$ and the action $A[\psi]$ are defined as follows:

$$D\psi = \prod_x \left[-d\psi^1(x)d\psi^2(x)d\psi^3(x)d\psi^4(x)\right] \quad (10.8)$$

$$A[\psi] = -\frac{1}{2}\sum_x \overline{\psi}(x)\psi(x) + \frac{1}{2}\lambda \sum_{x,\alpha} \overline{\psi}(x+\alpha)\hat{p}_\alpha\psi(x) \quad (10.9)$$

The 'conjugated' variables $\overline{\psi}(x)$ are defined as follows:

$$\overline{\psi^\alpha} = \psi^\gamma(\hat{C}^{-1})^{\gamma\alpha} \quad (10.10)$$

where

$$\hat{C} = \begin{pmatrix} 0 & 1 & 1 & 1 \\ -1 & 0 & 1 & 1 \\ -1 & -1 & 0 & 1 \\ -1 & -1 & -1 & 0 \end{pmatrix}; \; \hat{C}^{-1} = \begin{pmatrix} 0 & -1 & 1 & -1 \\ 1 & 0 & -1 & 1 \\ -1 & 1 & 0 & -1 \\ 1 & -1 & 1 & 0 \end{pmatrix} \quad (10.11)$$

and the vector matrix $\hat{p}_\alpha$ in Eq. (10.9) is:

$$\hat{p}_\alpha = \left\{ \begin{pmatrix} 1 & 0 & 0 & 0 \\ 1 & 0 & 0 & 0 \\ 0 & 0 & 0 & 0 \\ -1 & 0 & 0 & 0 \end{pmatrix}, \begin{pmatrix} 0 & 1 & 0 & 0 \\ 0 & 1 & 0 & 0 \\ 0 & 1 & 0 & 0 \\ 0 & 0 & 0 & 0 \end{pmatrix}, \begin{pmatrix} 0 & 0 & 0 & 0 \\ 0 & 0 & 1 & 0 \\ 0 & 0 & 1 & 0 \\ 0 & 0 & 1 & 0 \end{pmatrix}, \begin{pmatrix} 0 & 0 & 0 & -1 \\ 0 & 0 & 0 & 0 \\ 0 & 0 & 0 & 1 \\ 0 & 0 & 0 & 1 \end{pmatrix} \right\} \quad (10.12)$$

More explicitly for the action $A[\psi]$, Eq. (10.9), one gets:

$$A[\psi] = -\frac{1}{2}\sum_x \psi(x)\hat{C}^{-1}\psi(x) + \frac{1}{2}\lambda\sum_{x,\alpha}\psi(x+\alpha)\hat{C}^{-1}\hat{p}_\alpha\psi(x)$$
$$\equiv \sum_x \left[\psi^3(x)\psi^1(x) + \psi^4(x)\psi^2(x) + \psi^1(x)\psi^2(x) + \psi^3(x)\psi^4(x) + \psi^2(x)\psi^3(x)\right.$$

$$+\psi^1(x)\psi^4(x)\Big] + \lambda \sum_x \Big[\psi^3(x+1)\psi^1(x) + \psi^4(x+2)\psi^2(x)\Big] \qquad (10.13)$$

Using the rules (10.5) and (10.6) one can easily check by direct calculations that the integration in (10.7) with the integration measure (10.8) reproduces the high-temperature expansion of the 2D Ising model partition function (10.4) with $\lambda = \tanh\beta$.

Let us consider the Green function:

$$G^{\alpha\beta}(x,x') = Z^{-1} \int D\psi \psi^\alpha(x) \overline{\psi}^\beta(x') \exp(A[\psi]) \qquad (10.14)$$

Simple (although slightly cumbersome) calculations yield:

$$G^{\alpha\beta}(x,x') = \lambda \sum_\gamma \Lambda^{\alpha\gamma} G^{\gamma\beta}(x-\gamma,x') + \delta_{x,x'}\delta^{\alpha\beta} \qquad (10.15)$$

where $\hat{\Lambda} \equiv \sum_\alpha \hat{p}_\alpha$:

$$\hat{\Lambda} = \begin{pmatrix} 1 & 1 & 0 & -1 \\ 1 & 1 & 1 & 0 \\ 0 & 1 & 1 & 1 \\ -1 & 0 & 1 & 1 \end{pmatrix} \qquad (10.16)$$

If we perform a Fourier transformation of Eq. (10.15), it acquires the following matrix form:

$$\hat{G}(k) = (\hat{1} - \lambda\hat{\Lambda}(k))^{-1} \qquad (10.17)$$

where

$$\hat{\Lambda}(k) = \sum_\alpha \exp\{-i\mathbf{k}\alpha\}\hat{p}_\alpha = \begin{pmatrix} \exp(-ik_1) & \exp(-ik_2) & 0 & -\exp(ik_2) \\ \exp(-ik_1) & \exp(-ik_2) & \exp(ik_1) & 0 \\ 0 & \exp(-ik_2) & \exp(ik_1) & \exp(ik_2) \\ -\exp(-ik_1) & 0 & \exp(ik_1) & \exp(ik_2) \end{pmatrix} \qquad (10.18)$$

It is obvious from Eq. (10.17) that, if one of the eigenvalues of the matrix $\lambda\hat{\Lambda}(k)$ becomes unity, it signals a singularity. To find this point we first put the space momentum $k = 0$ (which corresponds to the infinite spatial scale). The four-valued indices of the Green function $G^{\alpha\beta}$ are related to four possible directions on a square lattice. Therefore, the idea is to perform the Fourier transformation over these angular degrees of freedom. One can easily check that the matrix $\hat{\Lambda}(0)$, (Eq. (10.18)), is diagonalizing in the following representation:

$$\psi_{\pm 1/2} = \frac{1}{2}\begin{pmatrix} 1 \\ \exp(\pm i\frac{\pi}{4}) \\ \exp(\pm i\frac{\pi}{2}) \\ \exp(\pm i\frac{3\pi}{4}) \end{pmatrix}; \quad \psi_{\pm 3/2} = \frac{1}{2}\begin{pmatrix} 1 \\ \exp(\pm i\frac{3\pi}{4}) \\ \exp(\pm i\frac{3\pi}{2}) \\ \exp(\pm i\frac{9\pi}{4}) \end{pmatrix} \tag{10.19}$$

The transformation matrix from the initial representation to the angular momentum (or spinor) representation with the above basic vectors, has the form:

$$\hat{U} = \frac{1}{2}\begin{pmatrix} 1 & 1 & 1 & 1 \\ E & \overline{E} & E^3 & \overline{E}^3 \\ E^2 & \overline{E}^2 & E^6 & \overline{E}^6 \\ E^3 & \overline{E}^3 & E^9 & \overline{E}^9 \end{pmatrix}; \quad E = \exp\left(i\frac{\pi}{4}\right), \quad \overline{E} = \exp\left(-i\frac{\pi}{4}\right) \tag{10.20}$$

In this representation we get:

$$\lambda\hat{\Lambda}'(0) = \lambda\hat{U}^{-1}\hat{\Lambda}(0)\hat{U} = \lambda\begin{pmatrix} \sqrt{2}+1 & 0 & 0 & 0 \\ 0 & \sqrt{2}+1 & 0 & 0 \\ 0 & 0 & -\sqrt{2}+1 & 0 \\ 0 & 0 & 0 & -\sqrt{2}+1 \end{pmatrix} \tag{10.21}$$

There is a singularity in Eq. (10.17) (at $k \to 0$) when one of the eigenvalues of $\lambda\hat{\Lambda}'$ becomes unity. From Eq. (10.21) we can easily find the critical point of the 2D Ising model:

$$\lambda_c \equiv \tanh\beta_c = \frac{1}{\sqrt{2}+1} \tag{10.22}$$

Another important point that follows from these considerations is that for the critical fluctuations in the vicinity of the critical point only states $\psi_{\pm 1/2}$ are important. Indeed it is easily checked (see below) that the correlation radius for $\psi_{\pm 1/2}$ goes to infinity as $\lambda \to \lambda_c$, while the correlations for $\psi_{\pm 3/2}$ are confined to lattice sizes.

Now, to describe the critical long-range fluctuations, which are responsible for the singularities in the thermodynamical functions, we can expand Eq. (10.17) near the point $\lambda = \lambda_c$. Using the explicit expression (10.18), and retaining only the first powers of k and $(\lambda - \lambda_c)/\lambda_c$, one gets:

$$\hat{G}(k) \simeq \frac{2\lambda_c^2}{\Delta} \begin{pmatrix} \tau - ik_1 & \frac{\tau - ik_1 - ik_2}{\sqrt{2}} & -ik_2 & -\frac{\tau - ik_1 + ik_2}{\sqrt{2}} \\ \frac{\tau - ik_1 - ik_2}{\sqrt{2}} & \tau - ik_2 & \frac{\tau + ik_1 - ik_2}{\sqrt{2}} & ik_1 \\ -ik_2 & \frac{\tau + ik_1 - ik_2}{\sqrt{2}} & \tau + ik_1 & \frac{\tau + ik_1 + ik_2}{\sqrt{2}} \\ -\frac{\tau - ik_1 + ik_2}{\sqrt{2}} & ik_1 & \frac{\tau + ik_1 + ik_2}{\sqrt{2}} & \tau + ik_2 \end{pmatrix} \tag{10.23}$$

Here

$$\Delta = \det[\hat{1} - \lambda\hat{\Lambda}(k)] \simeq 2\lambda_c^2(\tau^2 + k^2) \tag{10.24}$$

and

$$\tau \equiv 2\frac{(\lambda - \lambda_c)}{\lambda_c} \tag{10.25}$$

In the spinor representation given by Eq. (10.19) the asymptotic expression for Eq. (10.23) simplifies to the following compact form:

$$\hat{G}_{sp}(k) = \hat{U}^{-1}\hat{G}(k)\hat{U} \simeq \frac{2}{\tau^2 + k^2} \begin{pmatrix} \tau & ik_1 - k_2 & 0 & 0 \\ ik_1 + k_2 & \tau & 0 & 0 \\ 0 & 0 & 0 & 0 \\ 0 & 0 & 0 & 0 \end{pmatrix} \tag{10.26}$$

The zero components here are $\sim k^2, \tau^2$. The non-zero 2×2 block can be represented as:

$$\hat{S}(k) = 2\frac{\tau + i\hat{k}}{\tau^2 + k^2} \tag{10.27}$$

Here

$$\hat{k} = k_1\hat{\gamma}_1 + k_2\hat{\gamma}_2; \tag{10.28}$$

$$\gamma_1 = \begin{pmatrix} 0 & 1 \\ 1 & 0 \end{pmatrix}, \gamma_2 = \begin{pmatrix} 0 & i \\ -i & 0 \end{pmatrix} \tag{10.29}$$

The result (10.27) is the Green function of the free (real) spinor field in two

Euclidian dimensions described by the Lagrangian:

$$A_{sp}[\psi] = -\frac{1}{4}\int d^2x[\overline{\psi}\hat{\partial}\psi + \tau\overline{\psi}\psi] \tag{10.30}$$

where $\overline{\psi_1} = \psi_2$ and $\overline{\psi_2} = -\psi_1$.

Using Eq. (10.30) one immediately finds the logarithmic singularity of the specific heat of the 2D Ising model:

$$Z \simeq \int D\psi \exp\{A_{sp}[\psi]\} \simeq \sqrt{\det(\tau + \hat{\partial})};$$

$$F \simeq -\ln Z \simeq -\mathrm{Tr}\ln(\tau + \hat{\partial}) \simeq -\int d^2k\, \ln(\tau^2 + k^2) \sim -\tau^2 \ln\frac{1}{|\tau|} \tag{10.31}$$

Hence

$$C \sim -\frac{d^2}{d\tau^2}F(\tau) \sim \ln\frac{1}{|\tau|} \tag{10.32}$$

10.3 Critical behavior in the disordered model

We turn now to the model with disorder. The partition function of the 2D disordered Ising model is given by:

$$Z(\beta) = \sum_{\sigma} \exp\left(\beta \sum_{x,\mu} J_{x\mu}\sigma_x\sigma_{x+\mu}\right) \tag{10.33}$$

where the coupling constant $J_{x\mu}$ on a particular lattice bond (x, μ) is equal to the regular value J with probability $(1 - c)$, and to the impurity value $J' \neq J$ with probability c. We impose no restriction on J' but we shall require $c \ll 1$, so that the concentration of impurities is assumed to be small.

The Grassmann variables technique described in the previous section can be applied to the model with random lattice couplings as well. In this representation the partition function (10.33) is given by:

$$Z(\beta) = \int D\psi \exp\left[-\frac{1}{2}\sum_x \overline{\psi}(x)\psi(x) + \frac{1}{2}\sum_{x,\mu} \lambda_{x\mu}\overline{\psi}(x+\mu)\hat{P}_\mu\psi(x)\right] \tag{10.34}$$

where

$$\lambda_{x\mu} = \begin{cases} \lambda \;= \tanh(\beta J) & \text{with probability } (1-c) \\ \lambda' = \tanh(\beta J') & \text{with probability } c \end{cases} \tag{10.35}$$

It is easy to check by direct expansion in powers of the second term in (10.34) that the partition function can be represented as a sum over configurations of closed loops, each loop entering with a weight

$$\prod \lambda_{x\mu}\Phi(\mathscr{P}) \tag{10.36}$$

where $\Phi(\mathscr{P})$ is an ordered product along the path $\mathscr{P}$ of matrices $\{\hat{p}\}$:

$$\Phi(\mathscr{P}) = \prod_{\mathscr{P}} \hat{p} \tag{10.37}$$

The same representation for the partition function comes from the high-temperature expansion of Eq. (10.33).

Proceeding along these lines and averaging over the disorder in the couplings one could finally obtain the exact continuum limit representation for the free energy of the impurity model (see [48]). Here, however, we shall consider a more intuitive and much more simplified approach, which, nevertheless, provides the same results as the exact one. This approach is based on the natural assumption that in the continuum limit representation in terms of the free fermion fields (see previous section) the disorder in the couplings manifests itself as a small spatial disorder in the effective temperature τ of spinor Lagrangian (10.30). Therefore, the starting point for further considerations of the disordered model will be the assumption that its continuum limit representation is described by the following spinor Lagrangian:

$$A_{\text{imp}}[\psi;\delta\tau(x)] = -\frac{1}{4}\int d^2x \left[\overline{\psi}\hat{\partial}\psi + (\tau + \delta\tau(x))\overline{\psi}\psi\right] \tag{10.38}$$

Here the quenched random variable $\delta\tau(x)$ is assumed to be described by simple Gaussian distribution:

$$P[\delta\tau(x)] = \prod_x \left[\frac{1}{\sqrt{8\pi u}} \exp\left(-\frac{(\delta\tau(x))^2}{8u}\right)\right] \tag{10.39}$$

where the small parameter $u \ll 1$ is proportional to the concentration of impurities.

Then, the selfaveraging free energy can be obtained in terms of the traditional replica approach (Section 1.3):

$$F \equiv \overline{F[\delta\tau(x)]} = -\frac{1}{\beta}\lim_{n\to 0}\frac{1}{n}\ln(Z_n) \tag{10.40}$$

where

$$Z_n = \int D\delta\tau(x) P[\delta\tau(x)] \times \int D\psi^a \exp\left(-\frac{1}{4}\int d^2x \sum_{a=1}^{n}\left[\overline{\psi^a}\hat{\partial}\psi^a + (\tau + \delta\tau(x))\overline{\psi^a}\psi^a\right]\right) \tag{10.41}$$

is the replica partition function and the superscript $a = 1, 2, \ldots, n$ denotes the replicas. Simple Gaussian integration over $\delta\tau(x)$ yields:

$$Z_n = \int D\psi^a \exp\{A_n[\psi]\} \tag{10.42}$$

where

$$A_n[\psi] = -\int d^2x \left[\frac{1}{4}\sum_{a=1}^{n} \overline{\psi^a}(\hat{\partial}+\tau)\psi^a - \frac{1}{4}u\sum_{a,b=1}^{n} \overline{\psi^a}\psi^a\overline{\psi^b}\psi^b\right] \tag{10.43}$$

Note that rigorous perturbative consideration of the original lattice problem [48] yields the same result for the continuous limit effective Lagrangian (10.43), in which

$$u = c\frac{(\frac{\lambda_c'-\lambda_c}{\lambda_c})^2}{(1+\frac{1}{2\sqrt{2}}(\lambda_c'-\lambda_c))^2} \tag{10.44}$$

where

$$\lambda_c = \tanh\beta_c J = \sqrt{2}-1;$$
$$\lambda_c' = \tanh\beta_c J' \tag{10.45}$$

The spinor-field theory with the four-fermion interaction (10.43) obtained above is renormalizable in two dimensions, just as the vector field theory with the interaction ϕ^4 is renormalizable in four dimensions (Sections 7.5 and 8.3). Indeed, after the scale transformation (see Section 7.3):

$$x \to \lambda x \quad (\lambda > 1) \tag{10.46}$$

one gets:

$$\int d^D x\overline{\psi}(x)\hat{\partial}\psi(x) \to \lambda^{D-1}\int d^D x\overline{\psi}(\lambda x)\hat{\partial}\psi(\lambda x)$$

$$u\int d^D x\,(\overline{\psi}(x)\psi(x))\,(\overline{\psi}(x)\psi(x)) \to \lambda^D u\int d^D x\,(\overline{\psi}(\lambda x)\psi(\lambda x))\,(\overline{\psi}(\lambda x)\psi(\lambda x)) \tag{10.47}$$

To leave the gradient term of the Hamiltonian (which is responsible for the scaling of the correlation functions) unchanged, one has to rescale the fields:

$$\psi(\lambda x) \to \lambda^{-\Delta_\psi}\psi(x) \tag{10.48}$$

with

$$\Delta_\psi = \frac{D-1}{2} \tag{10.49}$$

The scale dimensions Δ_ψ defines the critical exponent of the correlation function:

$$G(x) = \langle\overline{\psi}(0)\psi(x)\rangle \sim |x|^{-2\Delta_\psi}|_{D=2} = |x|^{-1} \tag{10.50}$$

To leave the Hamiltonian (10.43) unchanged after these transformations one has to rescale the parameter u:

$$u \to \lambda^{-\Delta_u}u \tag{10.51}$$

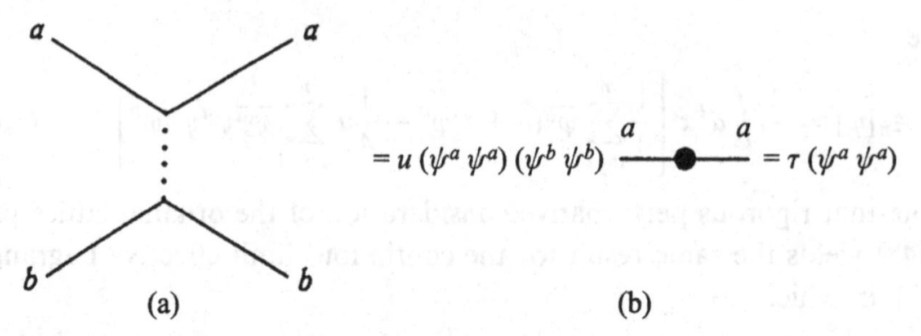

Fig. 10.2. Diagrammatic representation of the interaction term $u(\overline{\psi^a}(x)\psi^a(x))(\overline{\psi^b}(x)\psi^b(x))$ and the mass term $\tau(\overline{\psi^a}(x)\psi^a(x))$.

where

$$\Delta_u = 2 - D \tag{10.52}$$

Therefore, the scale dimension Δ_u of the four-fermion interaction term is zero in two dimensions, just as the scale dimension of the ϕ^4 interaction term is zero in four dimensions.

We shall see below that the renormalization equations lead to the 'zero-charge' asymptotics for the charge u and the mass τ. In this lucky case the critical behavior can be found by the renormalization-group methods or, in the same way, the main singularities of the thermodynamic functions can be found by summing up the 'parquette' diagrams of the theory (10.43) (cf. Section 7.5)

Let us perform the renormalization of the charge u and the mass τ. The diagrammatic representation of the interaction $u(\overline{\psi^a}(x)\psi^a(x))(\overline{\psi^b}(x)\psi^b(x))$ and the mass $\tau(\overline{\psi^a}(x)\psi^a(x))$ terms are shown in Fig. 10.2. It should be stressed that the model under consideration is described in terms of *real* fermions, and although we are using (just for convenience) the notation of the conjugated fields $\overline{\psi}$ they are not independent variables. For that reason the fermion lines in the diagram representation are not 'directed'. Actually, the interaction term (Fig. 10.2) can be represented explicitly in terms of only one two-component fermion (anticommuting) field: $u\psi_1^a\psi_2^a\psi_1^b\psi_2^b$. Therefore, the diagonal in replica indices ($a = b$) interaction terms are identically equal to zero.

Proceeding in a similar way to the calculations of Section 8.2 one then finds that the renormalization of the parameter u is provided only by the diagram shown in Fig. 10.3(c), whereas the first two diagrams, Figs. 10.3(a) and (b), are identically equal to zero. For the same reason the renormalization of the mass term is provided only by the diagram shown in Fig. 10.4(b), while the diagram in Fig. 10.4(a) is zero. The internal lines in Figs. 10.3 and

10.4 represent the massless free fermion Green function (cf. Eqs. (10.27) and (10.28)):

$$\hat{S}_{ab} = -i\frac{\hat{k}}{k^2}\delta_{ab} \tag{10.53}$$

Taking into account corresponding combinatorial factors one easily obtains the following RG transformation for the scale-dependent interaction parameter $u(\lambda)$ and mass $\tau(\lambda)$:

$$u^{(R)}(\lambda) = u + 2(n-2)u^2 \int_{\lambda k_0<|k|<k_0} \frac{d^2k}{(2\pi)^2}\mathrm{Tr}\hat{S}^2(k) \tag{10.54}$$

$$\tau^{(R)}(\lambda) = \tau + 2(n-1)u\tau \int_{\lambda k_0<|k|<k_0} \frac{d^2k}{(2\pi)^2}\mathrm{Tr}\hat{S}^2(k) \tag{10.55}$$

Using Eq. (10.53) after simple integration one gets the following RG equations (in the limit $n \to 0$):

$$\frac{d}{d\xi}u(\xi) = -\frac{2}{\pi}u^2(\xi) \tag{10.56}$$

$$\frac{d}{d\xi}\ln(\tau(\xi)) = -\frac{1}{\pi}u(\xi) \tag{10.57}$$

where, as usual, $\xi \equiv \ln(1/\lambda)$ is the RG parameter. These equations can be easily solved and yield:

$$u(\xi) = \frac{u}{1+\frac{2u}{\pi}\xi} \tag{10.58}$$

$$\tau(\xi) = \frac{\tau}{(1+\frac{2u}{\pi}\xi)^{1/2}} \tag{10.59}$$

where $u \equiv u(\xi = 0)$ and $\tau \equiv \tau(\xi = 0)$. At large scales ($\xi \to \infty$)

$$u(\xi) \sim \frac{1}{\xi} \to 0; \quad \tau(\xi) \sim \frac{1}{\sqrt{\xi}} \to 0 \tag{10.60}$$

The critical behavior of a model with the 'zero-charge' renormalization can be studied exactly by the RG methods. In a standard way one obtains for the singular part of the specific heat (cf. Section 8.3):

$$\begin{aligned} C(\tau) &\simeq -\frac{1}{2}\int_{|k|>|\tau|} \frac{d^2k}{(2\pi)^2}\mathrm{Tr}\hat{S}^2(k)\left(\frac{\tau(k)}{\tau}\right)^2 \\ &= \frac{1}{4\pi}\int_{|k|>|\tau|} \frac{dk}{k}\left(\frac{\tau(k)}{\tau}\right)^2 \\ &= \frac{1}{4\pi}\int_{\xi<\ln(1/|\tau|)} d\xi \left(\frac{\tau(\xi)}{\tau}\right)^2 \end{aligned} \tag{10.61}$$

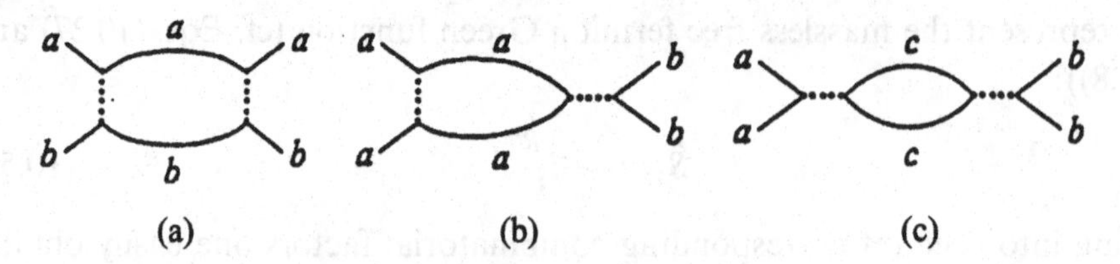

Fig. 10.3. The diagrams that contribute to the interaction term $u(\overline{\psi^a}(x)\psi^a(x))(\overline{\psi^b}(x)\psi^b(x))$.

Fig. 10.4. The diagrams that contribute to the mass term $\tau(\overline{\psi^a}(x)\psi^a(x))$.

Here the mass is taken to be dependent on the scale in accordance with Eq. (10.59):

$$\left(\frac{\tau(\xi)}{\tau}\right)^2 = \left(1 + \frac{2u}{\pi}\xi\right)^{-1} \tag{10.62}$$

Simple calculations yield:

$$C(\tau) \simeq \frac{1}{8u}\ln\left[1 + \frac{2u}{\pi}\ln\left(\frac{1}{|\tau|}\right)\right] \tag{10.63}$$

From (10.63) we see that in the temperature range $\tau_u \ll \tau \ll 1$ where

$$\tau_u \sim \exp\left(-\frac{\pi}{2u}\right) \tag{10.64}$$

the specific heat has the well known logarithmic behavior of the pure 2D Ising model: $C(\tau) \sim \ln(1/|\tau|)$. However, in the close vicinity of the phase transition point, at $|\tau| \ll \tau_u$, the specific heat exhibits different (universal) behavior:

$$C(\tau) \sim \frac{1}{u}\ln\left[\ln\left(\frac{1}{|\tau|}\right)\right] \tag{10.65}$$

which is still singular, although the singularity is now weaker.

Note that the critical exponent of the two-point correlation function in the 2D Ising model is not modified by the presence of disorder [51]:

$$\overline{\langle\sigma_0\sigma_x\rangle} \sim |x|^{-1/4} \tag{10.66}$$

This result is also convincingly confirmed by recent numerical simulations [55–57].

Note, finally, that the effects of the replica symmetry breaking (Chapter 9) in the present case appear to be irrelevant [52]. The corresponding calculations, although straightforward, are rather cumbersome and we do not reproduce them here. On the other hand, in the 2D Potts systems the disorder-induced RSB effects can be shown to be relevant and provide the existence of a non-trivial stable fixed point with continuous RSB (for details see [53]).

10.4 Numerical simulations

In recent years extensive numerical investigations on special purpose computers [54] have been performed, with the aim of checking the theoretical results derived for the 2D Ising model with impurity bonds [55, 56, 57]. In these studies, the calculations were performed for the model defined on a square lattice of $L \times L$ spins with the Hamiltonian

$$H = -\sum_{\langle i,j \rangle} J_{ij}\sigma_i\sigma_j \tag{10.67}$$

where the nearest neighbor ferromagnetic couplings J_{ij} are independent random variables taking two values J and J' with probabilities $1-u$ and u correspondingly.

Because the critical behavior of the disordered system is believed to be universal and independent of the concentration of impurities, it is much more convenient in numerical experiments to take the concentration u to be non-small. The point is that according to the theory discussed in the previous section, the parameter u defines the temperature scale $\tau_*(u)$ and correspondingly the spatial scale $L_*(u) \sim \exp\{\text{const}/u\}$, (Eq. (10.64)), at which the crossover to the disorder-induced critical behavior takes place. At small concentrations, the crossover scale L_* is exponentially large and it becomes inaccessible in numerical experiments for finite systems. On the other hand, if both coupling constants J and J' are ferromagnetic, then even for a finite concentration of impurity bonds the ferromagnetic ground state (and the ferromagnetic phase transition) is not destroyed, whereas the crossover scale L_* can be expected not to be very large.

Here we shall review only one set of numerical studies in which quite convincing results for the specific heat singularity have been obtained [56]. The model with the concentration of the impurities $u = 1/2$ has been studied. In this particular case the model given by Eq. (10.67) appears to be selfdual, and its critical temperature can be determined exactly from the equation [58]:

$$\tanh(\beta_c J) = \exp(-2\beta_c J') \tag{10.68}$$

In the Monte Carlo simulations a cluster-flip algorithm of Swendsen and Wang [59] was used; this algorithm overcomes the difficulty of critical slowing down. In one Monte Carlo sweep, the spin configuration is decomposed into clusters constructed stochastically by connecting neighboring spins of equal sign with the probability $(1 - \exp\{-2\beta J_{ij}\})$. Each cluster is then flipped with probability 1/2. At T_c and for large lattices, the relaxation to equilibrium for this algorithm appears to be much faster than for the standard single-spin-flip dynamics.

Technically it is much more convenient to calculate the maximum value of the specific heat as the function of the size of the system, instead of the direct dependence of the specific heat from the reduced temperature τ. Because the temperature and the spatial scales are in one to one correspondence ($R_c(\tau) \sim \tau^{-1}$ in the 2D Ising model), the minimum possible value for τ in a finite system of the size L is $\tau_{\min} \sim L^{-1}$. Therefore, the maximum value of the specific heat in the system that exhibits the critical behavior $C(\tau)$ must be of the order of $C(L^{-1})$. Then, according to Eq. (10.63), the size dependence of the specific heat in the disorder-induced critical regime, in the case of the 2D Ising model, can be expected to be as follows:

$$C(L) = C_0 + C_1 \ln(1 + b\ln(L)) \tag{10.69}$$

where C_0 and C_1 are some constants, and $b = 1/\ln(L_*)$, where L_* is the finite size impurity crossover length.

In general terms, the calculation procedure is as follows. First, one calculates the energy:

$$\overline{\langle H \rangle} = -\frac{1}{L^2}\overline{\left(\sum_{\langle i,j \rangle} J_{ij}\langle \sigma_i \sigma_j \rangle\right)} \tag{10.70}$$

where $\langle \ldots \rangle$ denotes the thermal (Monte Carlo) average. Then the specific heat is obtained from the energy fluctuations:

$$C(L) = L^2\overline{(\langle H^2 \rangle - \langle H \rangle^2)} \tag{10.71}$$

The simulations were performed for various ratios $r = J'/J = 1/10, 1/4, 1/2$ and 1. The system sizes ranged up to 600×600. Figure 10.5 displays the data for the critical specific heat, as determined from Eq. (10.71) at $r = 1/10, 1/4, 1/2$ and 1, plotted against the logarithm of L. For the sake of clarity, the vertical axis has been scaled differently for various r.

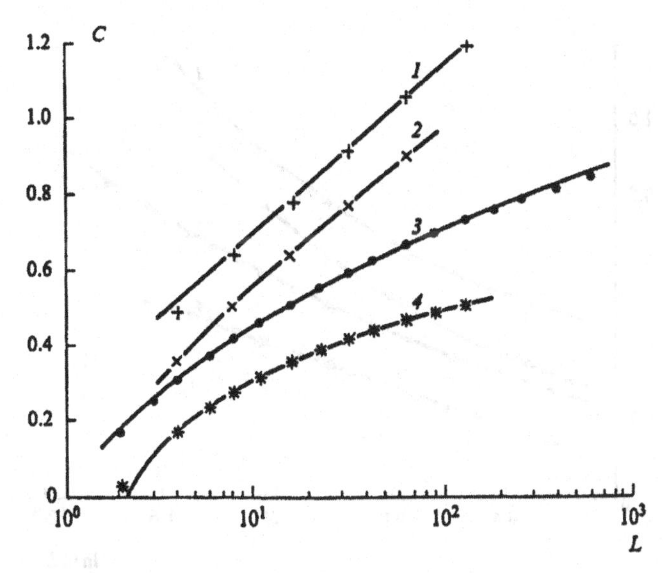

Fig. 10.5. The specific heat C at the critical temperature plotted as a function of $\ln L$: (*1*) the exact asymptotic result for the pure system, $r = 1$; (*2*) $r = 1/2$ with fitting parameters $C_0 = 0.048, C_1 = 15.7, b = 0.085$; (*3*) $r = 1/4$ with fitting parameters $C_0 = 0.048, C_1 = 2.04, b = 0.35$; (*4*) $r = 1/10$ with fitting parameters $C_0 = -0.28, C_1 = 0.224, b = 8.8$.

For the perfect model, $r = 1$, the deviations from the exactly known asymptotic behavior are obviously rather small for $L \geq 16$, in agreement with the analytic results on the corrections to scaling [60]. At $r = 1/2$ the size dependence data for $L \leq 128$ are still in the perfect Ising regime, where $C \sim \ln(L)$. At $r = 1/4$ and $r = 1/10$ strong deviations from the logarithmic size dependence occur, reflecting the crossover to the randomness-dominated region for sufficiently large values of L.

In Fig. 10.6 the same data are shown plotted against $\ln\ln(L)$. Strong upwards curvature is evident for $r = 1$ and $1/2$, indicating the logarithmic increase. In notable contrast, the data for $r = 1/4$ approach a straight line for moderate values of L, and those for $r = 1/10$ seems to satisfy such behavior even for small sizes, $L \geq 4$. From fits to Eq. (10.69), one obtains $L_* = 16 \pm 4$ at $r = 1/4$ and $L_* = 2 \pm 1$ at $r = 1/10$. The general trends are certainly clear, and confirm the expected crossover to a doubly logarithmic increase of C in the randomness-dominated region sets for smaller sizes L_* as r decreases.

Finally, in Fig. 10.7 the same data for $r = 1/4$ are plotted against $\ln(1 + b\ln(L))$, and exhibit a perfectly straight line for all values of L.

Therefore, in accordance with the analytical predictions of the renormal-

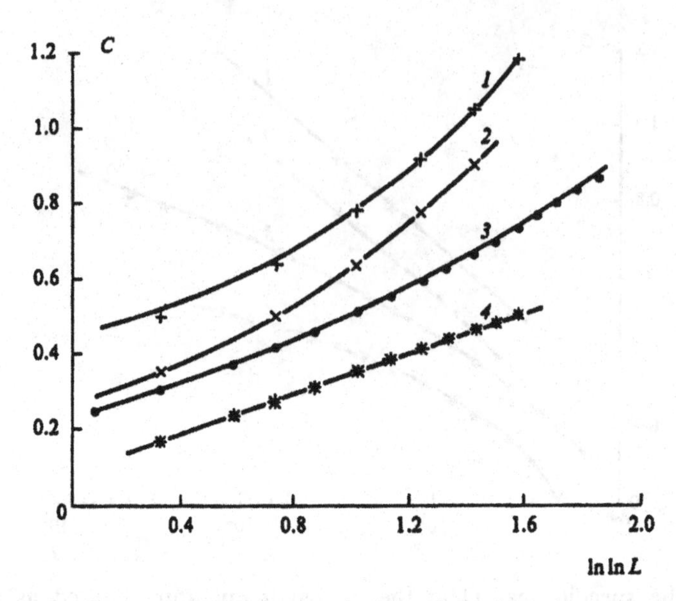

Fig. 10.6. The same set of data as in Fig. 10.5, plotted against $\ln \ln L$.

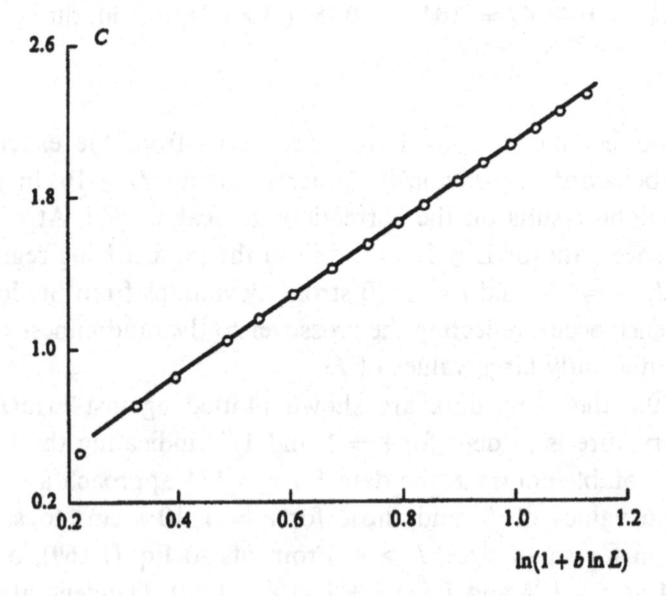

Fig. 10.7. The same set of data as in Fig. 10.5 for $r = 1/4$, plotted against $\ln(1+b\ln L)$ with $b = 0.35$.

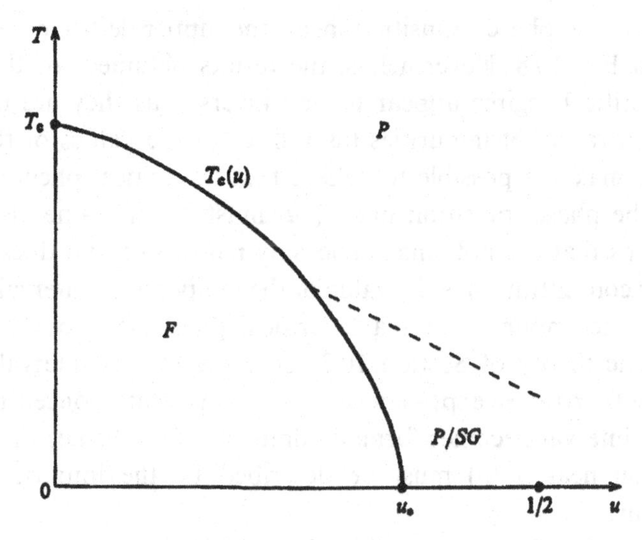

Fig. 10.8. A naive phase diagram of a ferromagnetic system diluted by antiferromagnetic or broken couplings.

ization-group calculations (Section 10.3), the results obtained in the numerical simulations provide convincing evidence for the onset of a new randomness-dominated critical regime. Besides, evidence is provided for a $\ln\ln(L)$ dependence in the behavior of the specific heat at the critical point for sufficiently large system sizes.

10.5 General structure of the phase diagram

Let us consider a general structure of the phase diagram of the Ising spin systems with impurities. Apparently, in a ferromagnetic system with antiferromagnetic or broken impurity bonds, as the concentration u of impurities increases, the ferromagnetic phase-transition temperature $T_c(u)$ decreases. Then, at some finite concentration u_c the ferromagnetic ground state can be completely destroyed, and correspondingly the phase-transition temperature should turn to zero: $T_c(u_c) = 0$. On the basis of these general arguments, one could guess that the qualitative phase diagram of such systems looks like that shown in Fig. 10.8 (for details, see e.g. [61], [62]). To the right of the line $T_c(u)$, the system is either in the paramagnetic state (at high enough temperatures) or in the spin-glass state [63]. The second possibility depends, however, on the dimensionality of the system; at $D = 2$ the spin-glass state is believed to be unstable at any non-zero temperature [64].

The critical phenomena considered in Section 10.3 formally correspond to the limit of small concentrations of impurities, i.e. they describe the properties of the phase transition near the upper left-hand side of the line $T_c(u)$ in Fig. 10.8. Nevertheless, the results obtained for the impurity-dominated critical regime appear to be universal, as they are independent of the concentration of impurities (as well as of the values of the impurity bonds). This makes it possible to believe that the critical phenomena in the vicinity of the phase-transition line $T_c(u)$ must be the same also for other concentrations that are not small. The only parameter that does depend on the impurity concentration is the value of the temperature interval near $T_c(u)$, $\tau_*(u)$, where the impurity-dominated critical phenomena occur. According to the analytic theory of Section 10.3 the value of this interval shrinks to zero as $u \to 0$: $\tau_*(u) \sim \exp\{-\text{const}/u\} \to 0$. At finite concentrations, this temperature interval becomes formally finite, which indicates that the whole critical region near $T_c(u)$ must be described by the impurity-dominated critical regime.

On the other hand, it is generally believed [61] that the bottom-right part of the phase-transition line $T_c(u)$ (the region near the critical concentration $u = u_c$, $T \ll 1$) belongs to another universality class, which is different from the ferromagnetic phase transition at $u \ll 1$. For example, it is obvious that in magnets with broken impurity bonds the phase transition as a function of the concentration (at $T \ll 1$) at $u = u_c$ must be of the kind of the percolation transition that has nothing to do with the ferromagnetic transition. It means that there must be a special point (T^*, u^*) on the line $T_c(u)$ which separates two different critical regimes.

Actually, there does exist a special line, the so-called Nishimori line $T_N(u)$ [65], which crosses the line $T_c(u)$ at the point (T^*, c^*) (Fig. 10.9). There is no phase transition at the Nishimori line. Formally it is special only in the sense that everywhere on this line the free energy as well as some other thermodynamic quantities appear to be analytic functions of the temperature and the concentration. Moreover, an explicit expression for free energy on the Nishimori line can be obtained for arbitrary T and u at any dimensions. In fact, it makes the structure of the phase diagram much less trivial than that shown in Fig. 10.8. Let us consider this point in more detail.

For the sake of simplicity, let us consider the Ising ferromagnet

$$H = -\sum_{\langle i,j \rangle} J_{ij} \sigma_i \sigma_j \tag{10.72}$$

defined at a lattice with arbitrary structure, where the ferromagnetic spin–spin couplings J_{ij} are equal to 1, while the impurity antiferromagnetic couplings are equal to -1, so that the statistical distribution of the J_{ij}s can be defined

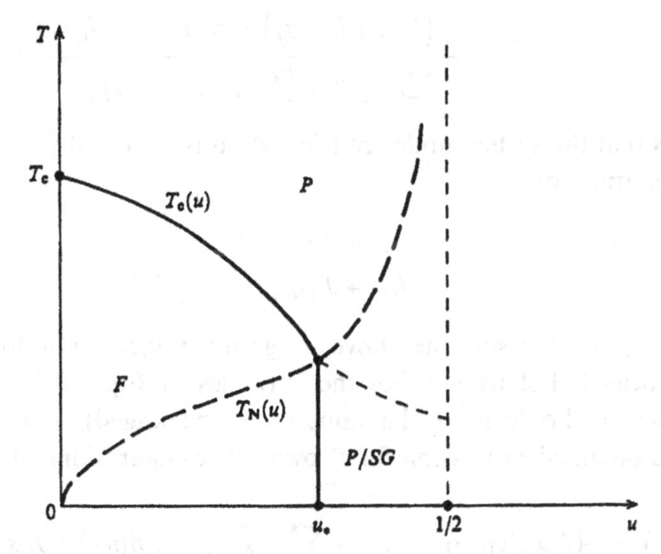

Fig. 10.9. Phase diagram of the Ising ferromagnet diluted by antiferromagnetic couplings; $T_{\rm N}(u)$ is the Nishimori line.

as follows:

$$P[J_{ij}] = \prod_{\langle i,j\rangle} [(1-u)\delta(J_{ij}-1) + u\delta(J_{ij}+1)] \tag{10.73}$$

where u is the concentration of the impurity bonds. One can easily check that the statistical averaging over configurations of the J_{ij}s:

$$\overline{(\ldots)} = \sum_{J_{ij}=\pm1} \prod_{\langle i,j\rangle} [(1-u)\delta(J_{ij}-1) + u\delta(J_{ij}+1)]\,(\ldots) \tag{10.74}$$

can be rewritten as follows:

$$\overline{(\ldots)} = \sum_{J_{ij}=\pm1} (2\cosh\tilde{\beta}(u))^{-N_{\rm b}} \exp\left(\tilde{\beta}(u)\sum_{\langle i,j\rangle} J_{ij}\right)(\ldots) \tag{10.75}$$

where $N_{\rm b}$ is the total number of bonds in the system, and the impurity parameter $\tilde{\beta}(u)$ is defined by the equation:

$$\exp\{-2\tilde{\beta}(u)\} = \frac{u}{1-u} \tag{10.76}$$

For given values of the temperature T and the concentration u the average energy of the system, $E(u,T) \equiv \overline{\langle H\rangle}$, can be represented as follows:

$$E(u,T) = -(2\cosh\tilde{\beta}(u))^{-N_{\rm b}} \sum_{J_{ij}=\pm1} \exp\left(\tilde{\beta}(u)\sum_{\langle i,j\rangle} J_{ij}\right)$$

$$\times \frac{\sum_{\sigma=\pm 1} \left(\sum_{\langle i,j \rangle} J_{ij}\sigma_i\sigma_j\right) \exp\left(\beta \sum_{\langle i,j \rangle} J_{ij}\sigma_i\sigma_j\right)}{\sum_{\sigma=\pm 1} \exp\left(\beta \sum_{\langle i,j \rangle} J_{ij}\sigma_i\sigma_j\right)} \tag{10.77}$$

It is obvious that the system under consideration is invariant under the local 'gauge' transformations:

$$\begin{aligned} \sigma_i &\to \sigma_i s_i \\ J_{ij} &\to J_{ij} s_i s_j \end{aligned} \tag{10.78}$$

for arbitrary $s_i = \pm 1$. Using the above gauge invariance the following trick can be performed. Let us redefine the variables in Eq. (10.77) according to (10.78) (which should leave the value of E unchanged), and then let us 'average' the obtained expression for E over all configurations of s_is:

$$\begin{aligned} E(u,T) = -(2\cosh\tilde{\beta}(u))^{-N_b} 2^{-N} \sum_{J_{ij}=\pm 1} \sum_{s=\pm 1} \exp\left(\tilde{\beta}(u) \sum_{\langle i,j \rangle} J_{ij} s_i s_j\right) \\ \times \frac{\sum_{\sigma=\pm 1} \left(\sum_{\langle i,j \rangle} J_{ij}\sigma_i\sigma_j\right) \exp\left(\beta \sum_{\langle i,j \rangle} J_{ij}\sigma_i\sigma_j\right)}{\sum_{\sigma=\pm 1} \exp\left(\beta \sum_{\langle i,j \rangle} J_{ij}\sigma_i\sigma_j\right)} \end{aligned} \tag{10.79}$$

One can easily see that the expression in Eq. (10.79)

$$\sum_{s=\pm 1} \exp\left(\tilde{\beta}(u) \sum_{\langle i,j \rangle} J_{ij} s_i s_j\right) \equiv Z[\tilde{\beta}(u), J_{ij}] \tag{10.80}$$

is the partition function of the system at the temperature $\tilde{\beta}(u)$. Therefore, if $\tilde{\beta}(u) = \beta$ the partition function (at the temperature β) in the denominator in the Eq. (10.79) is cancelled by the partition function (10.80). In this case the value of the average energy E (as well as the free energy) can be calculated explicitly:

$$\begin{aligned} E(u,T) &= -(2\cosh\tilde{\beta}(u))^{-N_b} 2^{-N} \sum_{J_{ij}=\pm 1} \sum_{\sigma=\pm 1} \left(\sum_{\langle i,j \rangle} J_{ij}\sigma_i\sigma_j\right) \exp\left(\beta \sum_{\langle i,j \rangle} J_{ij}\sigma_i\sigma_j\right) \\ &= -(2\cosh\tilde{\beta}(u))^{N_b} 2^{-N} \frac{\partial}{\partial\beta} \left[\sum_{J_{ij}=\pm 1} \sum_{\sigma=\pm 1} \exp\left(\beta \sum_{\langle i,j \rangle} J_{ij}\sigma_i\sigma_j\right)\right] \\ &= -N_b \tanh\beta(u) \\ &= -N_b(1 - 2u(T)) \end{aligned} \tag{10.81}$$

The internal energy obtained is analytic for all values of the temperature and the concentration.

The above result is valid at the Nishimori line $T_N(u)$ defined by the condition $\tilde{\beta}(u) = \beta$:

$$T_N(u) = \frac{2}{\ln \frac{1-u}{u}} \tag{10.82}$$

This line is shown qualitatively in Fig. 10.9. It starts for the zero concentration (pure system) at $T = 0$, and for $u \to 1/2$ (completely disordered system) $T_N \to \infty$. Apparently, the Nishimori line must cross the phase-transition line $T_c(u)$. This creates a rather peculiar situation, because at the line of the phase transition the thermodynamic functions should be non-analytic (for details, see [65]). Actually, this cross-section point, (T_*, u_*), is argued to be the multicritical point at which the paramagnetic, ferromagnetic and spin-glass phases merge [66].

For the Ising models of this type it can also be proved rigorously [65] that the ferromagnetic phase does not exist for $u > u_*$, where u_* is the point at which the Nishimori line crosses the boundary between the paramagnetic and the ferromagnetic phases $T_c(u)$ (Fig. 10.9). (It means that the structure of the naive phase diagram shown in Fig. 10.8, in general, is not quite correct.) To prove this statement let us consider the following two-point correlation function:

$$G(x) = \overline{\langle \sigma_0 \sigma_x \rangle_\beta} \tag{10.83}$$

where $\langle \ldots \rangle_\beta$ denotes the thermal average for a given temperature β. Using once again the above trick with the gauge transformation (10.78) for the correlation function (10.83) one gets:

$$\begin{aligned}
G(x) &= (2\cosh\tilde{\beta}(u))^{-N_b} \sum_{J_{ij}=\pm 1} \exp\left(\tilde{\beta}(u) \sum_{\langle i,j \rangle} J_{ij}\right) \\
&\quad \times \frac{\sum_{\sigma=\pm1} (\sigma_0\sigma_x) \exp\left(\beta \sum_{\langle i,j\rangle} J_{ij}\sigma_i\sigma_j\right)}{\sum_{\sigma=\pm1} \exp\left(\beta \sum_{\langle i,j\rangle} J_{ij}\sigma_i\sigma_j\right)} = (2\cosh\tilde{\beta}(u))^{-N_b} 2^{-N} \times \sum_{J_{ij}=\pm1} \\
&\quad \times \sum_{s=\pm1} (s_0 s_x) \exp\left(\tilde{\beta}(u) \sum_{\langle i,j\rangle} J_{ij} s_i s_j\right) \frac{\sum_{\sigma=\pm1} (\sigma_0\sigma_x) \exp\left(\beta \sum_{\langle i,j\rangle} J_{ij}\sigma_i\sigma_j\right)}{\sum_{\sigma=\pm1} \exp\left(\beta \sum_{\langle i,j\rangle} J_{ij}\sigma_i\sigma_j\right)} \\
&= (2\cosh\tilde{\beta}(u))^{-N_b} 2^{-N} \sum_{s'=\pm1} \sum_{J_{ij}=\pm1} \exp\left(\tilde{\beta}(u) \sum_{\langle i,j\rangle} J_{ij} s'_i s'_j\right) \\
&\quad \times \langle (s_0 s_x) \rangle_{\tilde{\beta}(u)} \langle (\sigma_0\sigma_x) \rangle_\beta = \overline{\langle s_0 s_x \rangle_{\tilde{\beta}(u)} \langle \sigma_0 \sigma_x \rangle_\beta}
\end{aligned} \tag{10.84}$$

Thus, the absolute value of the correlation function given by Eq. (10.83) satisfies the condition:

$$|G(x)| = |\overline{\langle\sigma_0\sigma_x\rangle_\beta}| \leq \overline{|\langle s_0 s_x\rangle_{\tilde{\beta}(u)}|} \tag{10.85}$$

because the absolute value of any Ising ($|\sigma| = 1$) correlation function does not exceed one. Therefore the absolute value of the two-point correlation function calculated at the temperature T and at the impurity concentration u does not exceed the average of the absolute value of the corresponding correlation function calculated at the Nishimori line at the same impurity concentration. This quantity in the long-range limit $|x| \to \infty$ vanishes if the corresponding point on the Nishimori line is in the paramagnetic phase, which takes place for all concentrations $u > u_*$. On the other hand, the value of the correlation function $G(x)$ in the limit $|x| \to \infty$ becomes the square of the ferromagnetic magnetization: $G(|x| \to \infty) = m^2(T, u)$. Thus, the above simple arguments prove that $m(T, u) \equiv 0$ for $u > u_*$.

Most probably, the boundary line between the ferromagnetic and non-ferromagnetic (spin-glass) phases is vertical to the concentration axis as in Fig. 10.9 [65], although the existence of the reentrant phenomena cannot in general be excluded.

Part three

Other types of disordered system

11

Ising systems with quenched random fields

11.1 The model

In the previous chapters we have considered the spin systems in which the quenched disorder was introduced in the form of random fluctuations in the spin–spin interactions. There exists another class of statistical models in which the disorder is present in the form of random magnetic fields. This type of disorder is essentially different from that with random interactions because external magnetic fields break the symmetry with respect to the change of the signs of spins.

In the most simple form, the random field spin systems can be qualitatively described in terms of the following Ising Hamiltonian:

$$H = -\sum_{<i\neq j>}^{N} \sigma_i \sigma_j - \sum_i h_i \sigma_i \tag{11.1}$$

where the Ising spins $\{\sigma_i = \pm 1\}$ are placed in the vertices of a D-dimensional lattice with the ferromagnetic interactions between the nearest neighbors, and quenched random fields $\{h_i\}$ are described by the symmetric Gaussian distribution:

$$P[h_i] = \prod_i^N \left[\frac{1}{\sqrt{2\pi h_0^2}} \exp\left(-\frac{h_i^2}{2h_0^2} \right) \right] ; \quad h_0 \ll 1 \tag{11.2}$$

The random field Ising model has numerous experimentally accessible realizations. Extensive experimental studies of the systems of this type have been performed using site diluted antiferromagnets in a homogeneous magnetic field [67], which in their statistical properties well reproduce the main features of the random field Ising model. On a qualitative level the equivalence of these two types of system can be understood as follows. An ordered antiferromagnetic system in the ground state is described by two sublattices,

A and B, with magnetizations being equal in their magnitude and opposite in signs. Dilution means that some spins, chosen at random, are removed from both sublattices. In zero external magnetic field the dilution alone does not break the symmetry between the two ground states $\sigma_A = -\sigma_B = \pm 1$. However, in a non-zero magnetic field h an isolated missing spin on the sublattice A provides the energy difference $2h$ between the two ground states $\sigma_A = -\sigma_B = +1$ and $\sigma_A = -\sigma_B = -1$, and it is this breaking of the symmetry that corresponds to the main feature of the random field Ising model.

Another example is absorbed monolayers with two ground states on impure substrates [68]. Here, if one of the substrate lattice sites is occupied by a quenched impurity it prevents additional occupation of this site, which effectively acts as a local symmetry breaking field. Other realizations are binary liquids in porous media [69], and diluted frustrated antiferromagnets [70].

11.2 General arguments

Despite extensive theoretical and experimental efforts during the past 20 years or so (for reviews see e.g. [15]) very little is understood about even the basic thermodynamic properties of the random field Ising model. According to simple physical arguments by Imry and Ma [71] it should be expected that the dimension D_c above which the ferromagnetic ground state is stable at low temperatures (it is called the lower critical dimension) must be $D_c = 2$. (Note, that for the Ising systems without random fields the lower critical dimension $D_c = 1$.) Indeed, if we try to test the stability of the ferromagnetic state by flipping the sign of the magnetization in a large region Ω_L of linear size L, we will find two competing effects: a possible gain of energy, $E_h(\Omega_L)$, due to alignments with the random magnetic fields, and the loss of energy, $E_f(L)$, due to the creation of an interface. The typical value of the (random) energy $E_h(\Omega_L)$, can be estimated as follows:

$$E_h \equiv \sqrt{\overline{E_h^2(\Omega_L)}} \sim \sqrt{\overline{(\sum_{i\in\Omega_L} h_i)^2}} = \sqrt{\sum_{i,j\in\Omega_L} \overline{h_i h_j}} \sim h_0 L^{D/2} \qquad (11.3)$$

The energy of the domain wall $E_f(L)$ is proportional to the square of the boundary of the region Ω_L:

$$E_f \sim L^{(D-1)} \qquad (11.4)$$

These estimates show that below dimensions $D_c = 2$ for any non-zero value of the field h_0 at sufficiently large sizes L the two energies are becoming comparable, and therefore no spontaneous magnetization should be present.

On the other hand, at dimensions greater than $D_c = 2$, the energy of the interface, E_f, is always bigger than E_h. Therefore these excitations will not destroy the long-range order, and a ferromagnetic transition should be present. These naive (but physically correct) arguments were later confirmed by a rigorous proof by Imbrie [72].

On the other hand, a perturbative study of the phase transition shows that, as far as the leading large-scale divergences are concerned, the strange phenomenon of a dimensional reduction is present, such that the critical exponents of the system in the dimension D appear to be the same as those of the ferromagnetic system without random fields in the dimension $d = D - 2$ [73]. Because the lower critical dimension of the pure Ising model is equal to 1, this result would imply that the lower critical dimension of the random field Ising model must be equal to 3, in contradiction with the rigorous results. Actually, the procedure of summation of the leading large-scale divergences could give the correct result only if the Hamiltonian in the presence of random fields has only one minimum. In this case the dimensional reduction can be rigorously shown to be exact, by the use of supersymmetric arguments [74].

However, as soon as the temperature is close enough to the critical point, as well as in a low-temperature region, there are values of the magnetic fields for which the free energy has more than one minimum (this phenomenon is similar to that considered in Chapter 9). In this situation there is no reason to believe that the supersymmetric approach should give the correct results and therefore the dimensional reduction is not grounded. This is not surprising, because the dimensional reduction completely misses the appearance of the Griffiths singularities [44].

Recently it has also been shown that the existence of more than one solution of the saddle-point equations in the presence of random fields is related, in the replica approach, to the existence of new instanton-type solutions of the mean-field equations which are not invariant under translations in the replica space [75, 88].

11.3 Griffiths phenomena in the low-temperature phase

It is well known both experimentally and by numerical simulations that although the ground state of the three-dimensional random field Ising system is ferromagnetic, if in the process of thermal equilibration the initial state of the system is taken to be disordered, in the low-temperature region it becomes trapped in metastable disordered states for macroscopically long times. Such states are characterized by large-scale ferromagnetically ordered domains with '+' or '−' magnetizations. This phenomenon clearly indicates that in the low-temperature region, in addition to the ferromagnetic ground state, the free energy must have many local minima states characterized by large-scale ordered spin domains. Besides strong dynamical slowing down, such well separated local minima could become the origin of the Griffiths singularities (exponentially small non-analytic contributions) in the thermodynamical functions.

In this section, by using simple physical arguments we shall demonstrate the origin of the Griffiths singularities in the low-temperature ferromagnetic phase of the random field Ising model in dimensions $D < 3$ [76]. In the low-temperature interval $h_0^2 \ll T \ll 1$ these non-perturbative contributions will be shown to come from rare, large spin clusters having characteristic sizes $\sim \sqrt{T}/h_0 \gg 1$ with a magnetization opposite to the ferromagnetic background, and which are the *local* minima of the free energy.

Because the ground state of the random field Ising system in dimensions greater than 2 is ferromagnetic, the thermal excitations are spin clusters with flipped magnetization. If the linear size L of such a cluster is large, then (in the continuous limit) the energy of this thermal excitation can be estimated as follows:

$$E(L) \simeq L^{D-1} - V(L) \tag{11.5}$$

where

$$V(L) = \int_{|x|<L} d^D x \, h(x) \tag{11.6}$$

The statistical distribution of the random function $V(L)$ (which is the energy of the spin cluster of the size L in the random field $h(x)$) is:

$$P[V(L)] = \int Dh(x) \exp\left(-\frac{1}{2h_0^2}\int d^D x h^2(x)\right) \prod_L \left[\delta\left(\int_{|x|<L} d^D x h(x) - V(L)\right)\right] \tag{11.7}$$

(here and in what follows all kinds of the pre-exponential factor are omitted). For future calculations it will be more convenient to deal with the quenched function $V(L)$ instead of $h(x)$. By a straightforward calculation one can easily derive the explicit expression for the distribution function $P[V(L)]$, (Eq. (11.7)) (for the sake of simplicity the parameter L at the intermediate stage of calculations is taken to be discrete):

$$P[V(L)] = \left(\prod_x \int_{-\infty}^{+\infty} dh(x)\right) \left(\prod_{i=1}^{\infty} \int_{-\infty}^{+\infty} d\xi_i\right)$$

$$\times \exp\left[-\frac{1}{2h_0^2}\int d^D x h^2(x) + i\sum_{i=1}^{\infty}\xi_i\left(\int_{|x|<L_i} d^D x h(x) - V(L_i)\right)\right]$$

$$= \left(\prod_{i=1}^{\infty}\int_{-\infty}^{+\infty} d\xi_i\right)\exp\left[-i\sum_{i=1}^{\infty}\xi_i V(L_i)\right]\left(\prod_{x}\int_{-\infty}^{+\infty} dh(x)\right)$$

$$\times \exp\left[-\frac{1}{2h_0^2}\int d^D x h^2(x) + i\sum_{i=1}^{\infty}\int_{L_i<|x|<L_{i+1}} d^D x h(x)\sum_{j=i}^{\infty}\xi_j\right]$$

$$= \left(\prod_{i=1}^{\infty}\int_{-\infty}^{+\infty} d\xi_i\right)\exp\left[-i\sum_{i=1}^{\infty}\xi_i V(L_i) - \frac{1}{2}h_0^2\sum_{i=1}^{\infty}(L_{i+1}^D - L_i^D)\left(\sum_{j=i}^{\infty}\xi_j\right)^2\right]$$

$$= \exp\left[-\frac{1}{2h_0^2}\sum_{i=1}^{\infty}\frac{[V(L_{i+1}) - V(L_i)]^2}{L_{i+1}^D - L_i^D}\right] \tag{11.8}$$

Taking L to be continuous again, one finally gets:

$$P[V(L)] \simeq \exp\left[-\frac{1}{2h_0^2}\int dL \frac{1}{L^{D-1}}\left(\frac{dV(L)}{dL}\right)^2\right] \tag{11.9}$$

Because the probability of the flips of big spin clusters is exponentially small, their contributions to the partition function can be assumed to be independent (in other words, it is assumed that such clusters are non-interacting, as they are very far from each other). Because at dimensions $D > 2$ the energy $E(L) = L^{D-1} - V(L)$ is on average a growing function of L, it would be reasonable to expect that the deep local minima (if any) of this function are well separated and the values of the energies at these minima increase with the size L. In this situation the sufficient condition for the existence of a minimum somewhere beyond a given size L is:

$$\frac{dV(L)}{dL} > (D-1)L^{D-2} \tag{11.10}$$

The probability $P_{\min}(L)$ that the above condition is satisfied at a unit length at the given size L can be easily estimated using the general distribution function (11.9). It is obtained by integrating $P[V(L)]$ over all functions $V(L)$ conditioned by $dV(L)/dL > (D-1)L^{D-2}$ (at the given value of L). It is clear, however, that with the exponential accuracy, the result of such an integration is defined by the lower bound $(D-1)L^{D-2}$ for the derivative $dV(L)/dL$. Therefore, substituting $dV(L)/dL = (D-1)L^{D-2}$ into Eq. (11.9), with the exponential accuracy one gets:

$$P_{\min}(L) \sim \exp\left[-\frac{(D-1)^2 L^{D-3}}{2h_0^2}\right] \tag{11.11}$$

On the other hand, one can also easily estimate the probability $P_L(V)$ that a cluster with a given size L has the energy of interaction with random fields equal to a given value V. According to Eq. (11.6): $\overline{V^2(L)} \simeq h_0^2 L^D$. Because the distribution $P_L(V)$ must be Gaussian, one immediately gets:

$$P_L(V) \sim \exp\left[-\frac{V^2}{2h_0^2 L^D}\right] \tag{11.12}$$

Note that this result can also be obtained by the direct integration of the general distribution function $P[V(L)]$, Eq. (11.9), over all 'trajectories' $V(L)$ constrained by $V(L) = V$ at the given length L.

Note now the important property of the random energy function $E(L)$, Eq. (11.5), which follows from Eqs. (11.11)–(11.12): although at dimensions $D > 2$ the function $E(L)$ on average increases with L, the probability of finding a local minimum of this function at dimensions $D < 3$, according to Eq. (11.11), also increases with L. It is the competition of these two effects that produces a non-trivial contribution to the free energy.

In the limit of low temperatures, $T \ll 1$, in addition to the usual (small-scale) perturbative fluctuations near the ordered ferromagnetic state, the separate contribution to the free energy, coming from rare (large-scale) local minima flipped spin clusters discussed above, can be estimated as follows:

$$\begin{aligned}\Delta F &\sim -T\int_1^{\infty} dL \int dV P_{\min}(L) P_L(V) \exp\left[-\beta(L^{D-1} - V)\right]\\ &= -T\int_1^{\infty} dL \int dV \exp\left[-\frac{V^2}{2h_0^2 L^D} - \frac{(D-1)^2 L^{D-3}}{2h_0^2} + \beta V - \beta L^{D-1}\right]\end{aligned} \tag{11.13}$$

One can easily check that at dimensions $D < 3$ in the region of parameters: $h_0 \ll 1$ and $h_0^2 \ll T \ll 1$, the expression in the exponential of the above integral, as a function of V and L, has the maximum at

$$L_* = (\beta h_0^2)^{-1/2}\sqrt{\frac{1}{2}(D-1)(3-D)} \gg 1 \tag{11.14}$$

$$V_* = (\beta h_0^2) L_*^D \gg (\beta h_0^2) \tag{11.15}$$

which is well separated from the 'trivial' region $L \sim 1; V \sim \beta h_0^2$. The contribution to the free energy coming from this saddle point is:

$$\Delta F \sim \exp\left[-\frac{\text{const}}{2h_0^2}(\beta h_0^2)^{\frac{3-D}{2}}\right] \tag{11.16}$$

where

$$\text{const} = \frac{1}{2}(D+1)(D-1)^{\frac{D-1}{2}}\left(\frac{2}{3-D}\right)^{\frac{3-D}{2}} \tag{11.17}$$

The result (11.16) demonstrates that in the limit $h_0 \ll 1$, in addition to the usual small-scale thermal fluctuations in the vicinity of the ordered state (which can be taken into account by the usual perturbation theory), there exist essentially non-perturbative large-scale thermal excitations which produce exponentially small non-analytic contributions to the thermodynamics. These excitations are flipped large-scale spin clusters which are the *local* energy minima. At low temperatures, such that $h_0^2 \ll T \ll 1$, the characteristic size of these clusters is $L_* \sim \sqrt{T}/h_0 \gg 1$.

The phenomenon discussed above, although it seems to produce a negligibly small contribution to the thermodynamical functions, must be extremely important for the dynamical relaxation processes. The large clusters with flipped magnetization being the local minima, are separated from the ground state by large energy barriers and can produce an essential slowing down of the relaxation (see e.g. [77]). According to Eqs. (11.14)–(11.15), the typical value of the energy barrier, which separates the flipped 'saddle-point' cluster from the ground state, must be of the order of $V_* \sim (\beta h_0^2)^{-(D-2)/2} \gg 1$. Correspondingly, the typical relaxation time needed to flip such a cluster must be exponentially large:

$$\tau(T) \sim \exp\left[\beta(\beta h_0^2)^{-\frac{D-2}{2}}\right] \gg 1 \qquad (11.18)$$

However, in order to predict the time asymptotics of the relaxation processes one needs to obtain the whole *spectrum* of the relaxation times, which can not be done based only on such simple saddle-point arguments.

Unfortunately, the above results cannot be directly applied to three-dimensional systems. The dimension $D = 3$ appears to be marginal for the considered phenomena (at dimensions $D > 3$ this type of non-perturbative effect is absent). The problem is that at $D = 3$ the above simple exponential estimate for the probability of local minima $P_{\min}(L)$, Eq. (11.11), makes no sense, and therefore to derive meaningful results much more detailed analysis is required.

On the other hand, it seems quite reasonable to expect that the results obtained here must remain correct also at dimensions $D = 2$ regardless of the fact that the long-range order in not stable there. The point is that at $D = 2$ the correlation length at which the long-range order is destroyed is believed to be exponentially large in the parameter $1/h_0$, whereas the typical scales, relevant for the discussed phenomena, are only the power of the parameter $1/h_0$, and therefore at such scales the system can be considered to be effectively ordered.

11.4 The phase transition

The nature of the phase transition in the random field Ising model is still a mystery. The only well established fact about it is that the upper critical dimensionality (the dimensionality above which the critical phenomena are described by the mean-field theory, Section 7.1) of such systems is equal to 6 (unlike pure systems, where it is equal to 4). Let us consider this point in more detail.

Near the phase transition the random field Ising model can be described in terms of following the scalar field Ginzburg–Landau Hamiltonian:

$$H = \int d^D x \left[\frac{1}{2}(\nabla\phi(x))^2 + \frac{1}{2}\tau\phi^2(x) - h(x)\phi(x) + \frac{1}{4}g\phi^4(x) \right] \tag{11.19}$$

where quenched random fields $h(x)$ are described by the usual δ-correlated symmetric Gaussian distribution with the mean square equal to h_0^2. Ground-state configurations of the fields $\phi(x)$ are defined by the saddle-point equation:

$$-\Delta\phi(x) + \tau\phi(x) + g\phi^3(x) = h(x) \tag{11.20}$$

In the usual RG approach for the phase transition in the pure systems ($h(x) = 0$) one constructs the perturbation theory over large-scale fluctuations on the background homogeneous solution of the above equation, $\phi_0 = \sqrt{|\tau|/g}$, $\tau < 0$ or $\phi_0 = 0$, $\tau > 0$ (Section 7.4). Apparently, the solutions of Eq. (11.20) with non-zero $h(x)$ essentially depend on a particular configuration of the quenched fields being non-homogeneous. Let us estimate the conditions under which the external fields become the dominant factor for the ground-state configurations.

Let us consider a large region Ω_L of a linear size $L \gg 1$. An average value of the field in this region can be defined as follows:

$$h(\Omega_L) = \frac{1}{L^D} \int_{x\in\Omega_L} d^D x h(x) \tag{11.21}$$

Correspondingly, for the typical value of these fields (averaged over realizations) one gets:

$$h_L \equiv \sqrt{\overline{h^2(\Omega_L)}} = \frac{1}{L^{2D}} \sqrt{\int_{x,x'\in\Omega_L} d^D x d^D x' \overline{h(x)h(x')}} = \frac{h_0}{L^{D/2}} \tag{11.22}$$

Then the estimate for the typical value of the order parameter field ϕ_L in this region can be obtained from the saddle-point equation:

$$\tau\phi_L + g\phi_L^3 = h_L \tag{11.23}$$

Then, as long as:

$$\tau\phi_L \ll g\phi_L^3 \tag{11.24}$$

the typical value of ϕ_L, according to Eq. (11.23), is defined only by the random field, and does not depends on τ:

$$\phi_L \sim \left(\frac{h_L}{g}\right)^{1/3} \sim \left(\frac{h_0}{g}\right)^{1/3} L^{-D/6} \tag{11.25}$$

Now let us estimate up to which characteristic sizes of the clusters the external fields can dominate. According to Eqs. (11.24) and (11.25) one gets:

$$L \ll \frac{(gh_0^2)^{1/D}}{\tau^{3/D}} \tag{11.26}$$

On the other hand, the estimation of the order parameter in terms of the equilibrium equation (11.23) can be correct only at scales much greater than the size of the fluctuation region, which is equal to the correlation length $R_c \sim \tau^{-\nu}$. Thus, one has the lower bound for L:

$$L \gg \tau^{-\nu} \tag{11.27}$$

Therefore, the region of parameters where the external fields dominate is:

$$\tau^{-\nu} \ll \frac{(gh_0^2)^{1/D}}{\tau^{3/D}} \tag{11.28}$$

or

$$\tau^{3-\nu D} \ll gh_0^2 \tag{11.29}$$

Such a region of temperatures near T_c exists only if:

$$\nu D < 3 \tag{11.30}$$

In this case the temperature interval near T_c in which the order parameter configurations are mainly defined by the random fields is:

$$\tau_c(h_0) \sim (gh_0^2)^{\frac{1}{3-\nu D}} \tag{11.31}$$

In the mean field theory (which correctly describes the phase transition in the pure system for $D > 4$) $\nu = 1/2$. Thus, according to the condition (11.30) the above non-trivial temperature interval τ_c exists only at dimensions $D < 6$. Substituting $\nu = 1/2$ into (11.31) we get:

$$\tau_c(h_0) \sim (gh_0^2)^{\frac{2}{6-D}} \tag{11.32}$$

Then, the random field defined spatial scale can be estimated from (11.26):

$$L_c(h_0) \sim (gh_0^2)^{\frac{-1}{6-D}} \tag{11.33}$$

Correspondingly, the typical value of the order parameter field at scales $L_c(h_0)$ is obtained from Eq. (11.25):

$$\phi_{L_c}^2 \sim \frac{1}{g}(gh_0^2)^{\frac{2}{6-D}} \tag{11.34}$$

Finally, the energy density can be estimated as $E/V \sim \phi_{L_c} h_{L_c}$. Taking into account (11.22) and (11.34) we find:

$$\frac{E}{V} \sim \frac{1}{g}(gh_0^2)^{\frac{4}{6-D}} \tag{11.35}$$

Note again that according to the above qualitative arguments all these non-trivial random field effects appear below upper critical dimensionality $D_c = 6$, while at dimensions $D > 6$ the phase transition must be correctly described in terms of the usual mean-field theory.

What is going on in the close vicinity of the phase transition point, $\tau \ll \tau_*(h_0)$, at dimensions $D < 6$, is not known. The main problem here is that in the studies of the critical properties one has to take into account that in the close vicinity of the critical point the saddle-point equation (11.20) can have many solutions [78]. The properties of a particular solution (whether it is a minimum or a maximum, the value of its energy etc.) essentially depend on a particular configuration of quenched random fields, and in such situation the usual RG approach, at least in its traditional form, can not be used.

Besides, according to recent numerical studies [79] there are indications of the existence of another 'critical' temperature T_* above T_c, such that at $T > T_*$ the solution of the saddle-point equation (11.20) is unique (this region corresponds to the usual paramagnetic phase), while at $T_c < T < T_*$ many saddle-point solutions appear, and finally only below T_c does the onset of the long-range magnetic order take place.

Although at the present state of knowledge in this field it is very difficult to hypothesize what could be the systematic approach to the problem, one of the possibilities is that the summation over numerous local minima states could be incorporated into the renormalization-group theory in the form of a particular replica symmetry breaking (RSB) structure of the parameters of the renormalized Hamiltonian. In this case one could face here new spin-glass-type critical phenomena similar to those discussed in Chapter 9.

It is already a long time since the possibility of a 'glassy' phase and the RSB phenomena in the random field Ising systems was first discussed [80]. Recently the RSB technique has been successfully applied for m-component ($m \gg 1$) spin systems with random fields [37], where it has been shown that in a finite temperature interval around the critical point the usual scaling replica-symmetric solution for spin–spin correlation functions is unstable with respect to the RSB. Moreover, a similar conclusion has also been derived for D-dimensional ($D < 6$) random field Ising systems, based on the double Legendre transform technique and general scaling arguments [81]. All this indicates that, owing to the existence of numerous local minima states, a special 'intermediate' (separating paramagnetic and ferromagnetic phase) spin-glass-like thermodynamic state could set in around the critical point.

12

One-dimensional directed polymers in random potentials

12.1 General scaling arguments

In a wide variety of physical systems one is interested in the behavior of a fluctuating linear object (with finite line tension) interacting with a quenched random potential. The object under consideration may be a dislocation in a crystal, a domain wall in a two-dimensional magnet, a vortex line in a superconductor, a fluxon line in an extended Josephson junction and so on, but following Ref. [82] this class of problem is traditionally discussed in terms of a directed polymer in random media or simply 'directed polymer'. The problem of a directed polymer in a random medium has been much studied in recent years [16].

Quite naturally the best understanding has been achieved for the simplest one-dimensional case when the displacements of a directed polymer can occur only in one direction. In such a case, a directed polymer in a continuous approximation can be described by the Hamiltonian

$$H[\phi(t), v] = \int_0^L dx \left\{ \frac{1}{2} (\nabla\phi)^2 + v[\phi(x), x] \right\} \tag{12.1}$$

where x is the longitudinal coordinate ($0 \leq x \leq L$) and $\phi(x)$ is the transverse displacement of a polymer with respect to a straight line. The simplest (or maybe one should better say the most easily treatable) assumption on the distribution of a random potential $v(\phi, x)$ consists of taking it to be Gaussian with

$$\overline{v(\phi, x)} = 0; \quad \overline{v(\phi, x) v(\phi', x')} = V(\phi - \phi')\delta(x - x') \tag{12.2}$$

Let us assume that at $x = 0$ the position of a polymer is fixed: $\phi(0) \equiv 0$. Then the quantity of interest is the typical deviation of the polymer 'trajectory' $\phi(x)$ from the origin. More precisely, one would like to know the dependence on L of the average square deviation of the polymer at the ending point $x = L$, which in the limit $L \to \infty$ is expected to be described by the simple scaling:

$$\overline{\langle \phi^2(L) \rangle} \equiv \overline{\left(Z_L^{-1} \int d\phi_0 \phi_0^2 \int_{\phi(0)=0}^{\phi(L)=\phi_0} D\phi(x) \exp(-\beta H[\phi(x), V]) \right)} \sim aL^{2\zeta} \tag{12.3}$$

where the partition function Z_L (for a given realization of the random potential) is given by the integration over all the trajectories $\phi(x)$ with only one boundary condition $\phi(x = 0) = 0$, and ζ is the so-called wandering exponent.

In the absence of the random potential the situation is trivial ($\langle x^2(L) \rangle \sim TL$) and the wandering exponent ζ is equal to $\frac{1}{2}$. In this case the trajectory deviates from the origin only because of the thermal fluctuations, the prefactor a in the scaling law (12.3) is proportional to the temperature, so at zero temperature $\langle x^2(L) \rangle = 0$.

In the presence of a quenched random potential the situation becomes much more complicated. Now besides the thermal fluctuations, the trajectory is pushed away from the origin also, owing to the randomness in the background potential landscape, so that the scaling law (12.3) could be governed by a new non-trivial wandering exponent. Moreover, because for a generic random potential the ground-state trajectory of the Hamiltonian (12.1) typically drifts away from the origin, the scaling law (12.3) can be expected to hold also in the zero temperature limit. The value of the wandering exponent ζ may essentially depend on the statistical properties of random potentials defined by the correlation function $V(\phi)$, Eq. (12.2).

One can consider two different types of statistics of random potentials. The case of the so-called *non-local* correlations can be desribed by a correlation function of the following type:

$$\overline{v(\phi, x) v(\phi', x')} = \delta(x - x') \left[(\text{const}) - g|\phi - \phi'|^{2\alpha} \right] \tag{12.4}$$

where $0 < \alpha < 2$. This problem naturally arises, with $\alpha = \frac{1}{2}$, when one considers an interface in the two-dimensional random field Ising model at low temperatures: then the field ϕ just describes the lateral fluctuations of the interface, in a solid-on-solid approximation.

In this case it is believed that the wandering exponent ζ should be equal to $3/2(2 - \alpha)$. A simple derivation of this scaling can be obtained by an energy balance argument *á la* Imry–Ma [71]. Let the value of the field be equal to ϕ_0 at $x = L$. Then the loss of the energy due to the gradient term in the Hamiltonian (12.1) can be estimated as $E_g \sim \phi_0^2/L$. The gain of energy due to the random potential term, according to Eq. (12.4), can be estimated as $E_V \sim -\sqrt{L}\sqrt{g}\phi_0^{\alpha}$. Optimizing E_g and E_V with respect to ϕ_0 one finds:

$$\phi_0 \sim L^{\frac{3}{2(2-\alpha)}}\, g^{\frac{1}{2(2-\alpha)}} \tag{12.5}$$

This result can be obtained in more rigorous terms from the Gaussian variational ansatz [83], and it can also be derived from a mapping to the Burgers (or the KPZ) equation and a study of this equation through a dynamical renormalization-group procedure [84]. In Section 13.2 we shall rederive this result in terms of the recently developed new method, which is called the vector replica symmetry breaking scheme.

A quite different result for the wandering exponent appears for the other class of random potentials, which are *locally correlated*, such that the function $V(\phi)$ in (12.2) quickly decays for $|\phi| \to \infty$. It is widely believed that in this case the wandering exponent ζ is *universal* and equal to 2/3. This conclusion is based on the finite-temperature exact results of Refs. [85, 86] and has been also confirmed by zero-temperature numerical simulations of the discrete version of the directed polymer problem [82] (see also Ref. [16] for later references). However, both the calculation based on the reduction to the damped Burgers equation with conservative random force [85], and the Bethe ansatz calculation in terms of replica representation [86] are valid only for the case of strictly δ-functional correlations of random potential.

In the next section we shall review some details of the approach developed by Kardar in Ref. [86] for the δ-correlated potential and use it to find the temperature dependence of a prefactor in Eq. (12.3), which turns out to be of the form $a \propto T^{-2/3}$. Then in Section 12.3 the generalization of this approach to a more physical situation when correlations of random potential have finite correlation radius r (in transverse direction) will be described. It will be demonstrated that in this case the usual Bethe ansatz replica solution is valid only in the high-temperature limit $T \gg T_0 \propto r^{2/3}$, whereas at low temperatures the solution has to have an essentially different structure. One can expect that if in the low-temperature limit the typical trajectory goes away from the origin, its drift should be determined by the quenched random potential and not by the effects of the thermal fluctuations. It means that in the system with reasonably short-scale *regularization* the divergence of the prefactor $a \propto T^{-2/3}$ at low temperatures can be expected to saturate. Indeed, it will be shown in Section 12.3 that the form of the low-temperature solution of the regularized problem can be described in terms of the effective one-step replica symmetry breaking ansatz. In this case one recovers the scaling law (12.3) with the same wandering exponent $\zeta = 2/3$ but with a *temperature-independent* prefactor [87].

12.2 Replica solution for δ-correlated potentials

The idea of Kardar's approach is based on the indirect calculation of the wandering exponent ζ by analyzing the scaling of the typical sample to sample *fluctuations* of the free energy. Suppose that the typical fluctuations of the free energy (produced by the random potential) scale as

$$\delta F \propto L^{\omega} \tag{12.6}$$

where the exponent ω is known. On the other hand, if the typical deviation of the trajectory from the origin is equal to ϕ, then the loss of the energy due to the elastic term in the Hamiltonian (12.1) must be of the order of ϕ^2/L. Balancing the two energies, one can write the following estimate:

$$\overline{\langle \phi^2 \rangle} \sim L\,\delta F \propto L^{\omega+1} \tag{12.7}$$

Then, according to the definition of the wandering exponent (12.3), one finds the following simple relation between the two exponents:

$$\zeta = \frac{1}{2}(\omega + 1) \tag{12.8}$$

The scaling of the free energy fluctuations with the size of the system L can be relatively easily investigated in terms of the replica method. As usual, the replica partition function

$$Z(n) \equiv \overline{Z^n[v]} \tag{12.9}$$

is given by the average of the n-th power of the partition function:

$$Z[v] = \int_{0<x<L} D\phi(x) \exp\{-\beta H[\phi(x), v]\} \tag{12.10}$$

obtained by the integration over all the trajectories with $\phi(0) = 0$.

According to the definition of the free energy: $F = -(1/\beta)\ln Z[v]$, the replica partition function $Z(n)$ can be represented as follows:

$$Z(n) = \overline{\exp[-\beta n F]} \tag{12.11}$$

On the other hand, the free energy $F \equiv F[v]$ is itself the sample-dependent random quantity, whose distribution function we shall denote as $P(F)$. Then Eq. (12.11) can be rewritten as

$$Z(n) = \int dF P(F) \exp[-\beta n F] \tag{12.12}$$

which is nothing else but the Laplace transform of the free energy distribution function $P(F)$.

It is natural to represent the replica partition function (12.9) in the following form:

$$Z(n) = \exp\{-\beta L E(n)\} \tag{12.13}$$

where the quantity $E(n)$ plays the role of the density of the replica free energy. Although this quantity can be calculated only for an integer n, according to the standard ideology of the replica approach it has to be considered as a function of the *continuous* parameter n, which implies the necessity of an analytic continuation in n. Comparing Eqs. (12.12) and (12.13) we get:

$$\exp\{-\beta L E(n)\} = \int dF P(F) \exp[-\beta n F] \tag{12.14}$$

Let us represent the density of the free energy $E(n)$ of the replicated system as a series in powers of the replica parameter n:

$$E(n) = \sum_{k=1}^{\infty} \frac{1}{k!} E_k n^k \tag{12.15}$$

Then, taking the k-th derivative over n at $n = 0$ from both sides of Eq. (12.14) for the k-th order of the free energy fluctuations one finds:

$$\beta L E_k = \beta^k \overline{\overline{F^k}} \tag{12.16}$$

where a double overbar denotes the irreduceable average. Thus the typical value of the k-th order free energy fluctuations, $\delta F_{(k)} \sim \left(\overline{\overline{F^k}}\right)^{1/k}$, scales with L as follows:

$$\delta F_{(k)} \sim \frac{1}{\beta} (\beta E_k)^{1/k} \, L^{1/k} \tag{12.17}$$

It will be shown below that in the case of the δ-correlated random potential, the series in powers of the replica parameter n, Eq. (12.15), contains only the linear (E_1) and the cubic (E_3) terms. The linear term yields the usual average free energy (which is linear in L), whereas the cubic term yields the scaling $L^{1/3}$ of the typical sample-to-sample free energy fluctuations. In other words, the exponent ω in Eq. (12.6) is equal to $1/3$, which according to Eq. (12.8) yields the value of the wandering exponent $\zeta = 2/3$. Let us consider this replica solution in some more detail.

In the framework of the replica approach the statistical mechanics of the *irregular* system is analyzed by considering the statistical mechanics of the *regular* system in which the disorder manifests itself in the form of the interaction between n identical replicas of the original system. For the system described by the Hamiltonian (12.1) and the Gaussian statistics of the random potential, Eq. (12.2), the averaging of $Z^n[v]$ over the disorder leads to the expression for $Z(n)$ the form of which corresponds to the following replica Hamiltonian:

$$-\beta H_n[\phi_a(x)] = \int_0^L dx \left\{ -\frac{1}{2}\beta \sum_{a=1}^{n} \left(\frac{d\phi_a}{dx}\right)^2 + \frac{1}{2}\beta^2 \sum_{a,b=1}^{n} V[\phi_a(x) - \phi_b(x)] \right\} \tag{12.18}$$

Let us redefine the replica partition function as follows:

$$Z(n) = \int_{-\infty}^{+\infty} dy_1 \dots dy_n Z_L(\mathbf{y}) \tag{12.19}$$

where $\mathbf{y} \equiv (y_1, y_2, \dots, y_n)$ is the n-component replica vector, and

$$Z_L(\mathbf{y}) = \int_{\phi_a(0)=0}^{\phi_a(L)=y_a} D\phi_a(x) \exp\{-\beta H_n[\phi_a(x)]\} \tag{12.20}$$

is the replica partition function with fixed boundary conditions at $x = L$. One can easily check that the partition function $Z_L(\mathbf{y})$ satisfies the following one-dimensional Shrödinger equation:

$$\frac{\partial}{\partial L} Z_L(\mathbf{y}) = \frac{1}{2\beta} \sum_{a=1}^{n} \frac{\partial^2}{\partial y_a^2} Z_L(\mathbf{y}) + \frac{1}{2}\beta^2 \sum_{a,b=1}^{n} V[y_a - y_b] Z_L(\mathbf{y}) \tag{12.21}$$

Thus, the Hamiltonian (12.18) corresponds to the Euclidean (imaginary time) action describing the quantum-mechanical system of n particles with mass β and interaction $\beta^2 V(y)$. The same system can be described by the quantum-mechanical (operator) Hamiltonian

$$\hat{H} = -\frac{1}{2\beta} \sum_{a=1}^{n} \nabla_a^2 - \frac{1}{2}\beta^2 \sum_{a,b=1}^{n} V(y_a - y_b) \tag{12.22}$$

which for the classical partition function defined by the Hamiltonion (12.18) plays the role of the transfer matrix.

In the limit of infinite size ($L \to \infty$) the free energy of a system (for any boundary conditions) is dominated by the highest eigenvalue of the transfer matrix or, in our case, by the lowest eigenvalue E_0 of the quantum-mechanical Hamiltonian (12.22). The stationary (in 'time' $L \to \infty$) solution of the quantum-mechanical problem (12.21) is given by the following wave function (partition function):

$$Z_L(\mathbf{y}) = \exp\{-\beta E(n) L\} \; \Psi(\mathbf{y}) \tag{12.23}$$

According to Eq. (12.21) we have:

$$-\beta E(n)\Psi(\mathbf{y}) = \frac{1}{2\beta}\sum_{a=1}^{n}\frac{\partial^2}{\partial y_a^2}\Psi(\mathbf{y}) + \frac{1}{2}\beta^2\sum_{a,b=1}^{n} V[y_a - y_b]\Psi(\mathbf{y}) \tag{12.24}$$

One can easily check that for any integer n the lowest eigenvalue corresponds to the fully symmetric wave function, which for the case of local correlations of the random potential (Eq. (12.2)):

$$V(y) = u\delta(y) \tag{12.25}$$

has the following simple form:

$$\Psi_0[\mathbf{y}] = \exp\left(-\frac{1}{2}\beta^3 u \sum_{a,b=1}^{n} |y_a - y_b|\right) \tag{12.26}$$

and the energy of this state is equal to

$$\beta E_0(n) = -\frac{1}{2}\beta^2 V(0) n - \frac{1}{6}\beta^5 u^2 n(n^2 - 1) \tag{12.27}$$

where the first term describes the trivial contribution to $E(n)$ related to the terms with $a = b$ in the second sum in the Hamiltonian (12.22). In terms of the representations (12.13),(12.15) this result reads:

$$E(n) = E_1 n + \frac{1}{6}E_3 n^3 \tag{12.28}$$

where

$$E_1 = -\frac{1}{2}\beta V(0) + \frac{1}{6}\beta^4 u^2 \tag{12.29}$$

and

$$E_3 = -\beta^4 u^2 \tag{12.30}$$

Comparison of Eq. (12.16) with Eq. (12.28) shows that for $L \to \infty$ the average free energy (per unit length) of a random polymer

$$f \equiv \lim_{L\to\infty}\frac{\overline{F(L)}}{L} \tag{12.31}$$

is given by the linear in n contribution to $E(n)$:

$$f = E_1 = -\frac{1}{2}\beta V(0) + \frac{1}{6}\beta^4 u^2 \tag{12.32}$$

in which the first (formally divergent) term always dominates. Therefore the average free energy of the system could be defined only after proper short-scale regularization of the starting Hamiltonian.

However, the fluctuations of the free energy are quite well defined without any regularization. According to Eqs. (12.17) and (12.30) the typical value of the free energy fluctuations can be estimated as:

$$\delta F \sim \frac{1}{\beta}(\beta E_3)^{1/3} L^{1/3} = \beta^{2/3} u^{2/3} L^{1/3} \tag{12.33}$$

Therefore, according to Eq. (12.7) for the average square deviation of the trajectory, one finds the following result:

$$\overline{\langle \phi^2 \rangle} \sim (\beta u)^{2/3} L^{4/3} \tag{12.34}$$

with the wandering exponent $\zeta = 2/3$.

12.3 Random potentials with finite correlation radius

Apparently the divergence of the average free energy, Eq. (12.32), is removed if one takes into account that in a physical system correlations of random potential should be described by a smooth function with a finite correlation radius (and therefore a finite value of $V(0)$). However, in such a case the quantum-mechanical problem defined by Eq. (12.21) cannot be solved exactly. Nonetheless it seems reasonable to assume that for narrow enough $V(\phi)$ one can still use the expression (12.25) in which u should now stand for

$$u = \int_{-\infty}^{+\infty} d\phi \, V(\phi) \tag{12.35}$$

It is easy to understand that such an approximate description (based on Eqs. (12.26)–(12.30)) at low temperatures has to fail. One has to remember that in the case of the δ-functional interaction the wave function (12.26) is constructed as a generalization of two-particle problem wave function

$$\Psi(y_1, y_2) = \exp(-æ|y_1 - y_2|) \tag{12.36}$$

On the other hand, in the case of a rectangular well:

$$V(\phi) = \begin{cases} V & \text{for } |\phi| < \frac{1}{2}r \\ 0 & \text{for } |\phi| > \frac{1}{2}r \end{cases} \tag{12.37}$$

(for which $u = Vr$) the wave function of the two-particle problem for $|y_1 - y_2| > r$ also has the form (12.36) with

$$æ = \frac{1}{r} g\left(2\beta^3 T_0^3\right) \tag{12.38}$$

where

$$g(z) \approx \begin{cases} z^{1/2} & \text{for } z \gg 1 \\ z & \text{for } z \ll 1 \end{cases} \tag{12.39}$$

and

$$T_0 = (Vr^2)^{1/3} \tag{12.40}$$

It is not hard to check that the condition $æ^{-1} \gg r$ for this wave function to be wide in comparison with the well width r coincides with the condition $T \gg T_0$. In such case the wave function (12.36) with the value of æ given by Eqs. (12.38)–(12.40) coincides with that of (the two-particle problem) (12.26). Correspondingly, the average free energy of the original problem of a directed polymer in the random potential described by the correlation function (12.37) (the replica n-particle system) is obtained from Eq. (12.32) by changing $u \to Vr$ and $V(0) = V$:

$$f = -\frac{1}{2}\beta V + \frac{1}{6}\beta^4 V^2 r^2 \tag{12.41}$$

The average square deviation of the trajectory is obtained from Eq. (12.34):

$$\overline{\langle\phi^2\rangle} \sim (\beta V r)^{2/3} L^{4/3} \tag{12.42}$$

In the opposite limit $T \ll T_0$ ($æ^{-1} \ll r$) the two-particle wave function is almost completely localized inside the well and cannot be used as a building block for the construction of the solution of the n-particle problem. It is not surprising, therefore, that the result (12.42) has no zero-temperature limit. This gives a clear indication that for the *arbitrary* finite-width form of the function $V(x)$ describing the correlations of random potential, the applicaton of Kardar's solution for the description of a random polymer can work only at high enough temperatures whereas in the low-temperature limit the solution should be different.

Quite paradoxically the differentiation of Eq. (12.41) shows that for $T \gg T_0$ the free energy defined by Eq. (12.41) corresponds to negative entropy. One should not be too scared of that property because in the approach discussed above the free energy of a directed polymer is calibrated in such a way that in the absence of the disorder it is equal to zero. Therefore the total free energy will be given by Eq. (12.41) *plus* the free energy in the absence of disorder. This second term will give the positive contribution to the entropy, which will overcome the negative contribution from Eq. (12.41).

Let us now consider the n-particle problem defined by Hamiltonian (12.22)

where

$$V(\phi) = \begin{cases} V(1-\phi^2) & \text{for } |\phi| < \frac{1}{2}r \\ 0 & \text{for } |\phi| > \frac{1}{2}r \end{cases} \tag{12.43}$$

Here, in comparison with Eq. (12.37), we have introduced a finite curvature of the potential inside the well. This parabolic well is truncated at the finite size described by the parameter r. At some stage of the calculations r will be assumed to be much smaller than one. For $V(\phi)$ of the form (12.43)

$$u \equiv \int_{-\infty}^{+\infty} d\phi V(\phi) = V\left(r - \frac{1}{12}r^3\right) \simeq Vr \tag{12.44}$$

and therefore the characteristic temperature T_0 defining the range of applicability of Kardar's approach ($T \gg T_0$) still can be chosen in the form (12.40).

It can be easily checked that for any *integer* number n of particles of the quantum-mechanical problem (12.24) in the limit of low temperatures $\beta \to \infty$, all particles are localized near the bottom of the well. Indeed, the *exact* ground-state solution of the Shrödinger equation (12.24) in which one assumes that the potential $V(\mathbf{y}) = V\sum_{a,b=1}^{k}[1-(y_a-y_b)^2]$ holds for all values of $(y_a - y_b)$, yields:

$$\Psi_0[\mathbf{y}] = \exp\left(-\frac{1}{4}\sqrt{\frac{2V\beta^3}{n}}\sum_{a,b=1}^{n}(y_a - y_b)^2\right) \tag{12.45}$$

and

$$\beta E_0(n) = -\frac{1}{2}V\beta^2 n^2 + \sqrt{\frac{1}{2}\beta n V}\,(n-1) \tag{12.46}$$

Using the explicit form of the wave function (12.45) one can easily estimate the typical value of the average distance between two arbitrary particles Δy in this n-particle system:

$$\Delta y \equiv \left[\frac{1}{n(n-1)}\sum_{a,b=1}^{n}(y_a - y_b)^2\right]^{1/2} \sim \frac{1}{(V\beta^3 n)^{1/4}} \tag{12.47}$$

In the limit $\beta \to \infty$ (at fixed integer value n), $\Delta y \to 0$. Therefore in the zero-temperature limit the wave function $\Psi(\mathbf{y})$ is indeed nicely localized at the bottom of the well for any integer n, and all corrections to $E_0(n)$ due to non-parabolicity can be only exponentially small.

On the other hand one can easily see why the limit $n \to 0$ is dangerous.

The width (12.47) of the wave function (12.45) grows with decreasing n and becomes comparable with the width of the well r at $n \sim T^3/Vr^4$, and therefore for smaller n the ground-state wave function should have an essentially different form. The simplest way to let the particles enjoy their mutual attraction while keeping their number in the well not too small consists of splitting them into n/k infinitely separated blocks of k particles. According to Eq. (12.46) the ground-state energy $E_0(n,k)$ of such a state with broken replica symmetry is given by

$$\beta E_0(n,k) = \frac{n}{k}\beta E_0(k) = \beta n \left[-\frac{1}{2}\beta k V + \sqrt{\frac{V}{2\beta k}}\,(k-1) \right] \tag{12.48}$$

and has the extremum (maximum) as a function of k. Variation of the polymer free energy per unit length:

$$f(k) = \frac{1}{n}E_0(n,k) = -\frac{1}{2}\beta k V - \sqrt{\frac{V}{2\beta k}}\,(1-k) \tag{12.49}$$

with respect to k gives the following equation for the position of the maximum:

$$-V + \sqrt{\frac{V}{2(\beta k)^3}}\,(1+k) = 0 \tag{12.50}$$

The solution of this equation in the zero-temperature limit, $\beta \to \infty$, is:

$$k_* \approx \frac{1}{\beta}(2V)^{-1/3} \tag{12.51}$$

Using this solution we can now estimate the value of the mean separation of particles belonging to the same block. According to Eq. (12.47), (where we have to substitute $n \to k_*$), for the mean separation of particles we find:

$$\Delta y \sim (V\beta^3 k_*)^{-1/4} \sim V^{-1/6}\beta^{-1/2} \tag{12.52}$$

Thus, at low enough temperatures, $T \ll V^{1/3}r^2$, the replicas belonging to the same block are indeed tightly bound to each other ($\Delta y \ll r$), so the whole picture is really selfconsistent.

Substituting the solution (12.51) into Eq. (12.49) for the free energy we find the following temperature-independent result:

$$f \approx -\frac{3}{4}(2V)^{2/3} \tag{12.53}$$

Although we have found that in the extremum solution the particles split into n/k separate blocks there are no reasons for these blocks to be infinitely separated from each other. The presence of strong attraction between the particles in each block makes it possible to consider such a block as a complex particle with the mass βk, the interaction between these complex particles being given by $\beta^2 k^2 V(x)$. The last expression can be expected to be very accurate when we consider the temperature interval $T \ll T_0$ in which the distances between the particles inside each block are much smaller than the well radius r.

Therefore at low temperatures the behavior of our system in which the particles are assumed to be tightly bound in n/k separate blocks can be described by the Hamiltonians (12.18) and (12.22) in which

$$\beta \to \beta k; \quad n \to \frac{n}{k} \tag{12.54}$$

Our earlier experience tells us that for some values of parameters such a system can be rather accurately described by the wave function of the form (12.26), in which y_a now stands for coordinates of different blocks. The energy of such a state will be given by Eq. (12.27) in which substitutions (12.54) and (12.44) have to be made with the first term being substituted by Eq. (12.48):

$$\beta E_0(n,k) = \beta n \left[-\frac{1}{2}\beta k V + \sqrt{\frac{V}{2\beta k}}\,(k-1) + \frac{1}{6}(\beta k)^4 V^2 r^2 \right] - \frac{1}{6}\beta^3 n^3 V^2 r^2 (\beta k)^2 \tag{12.55}$$

All this leads to the appearance in the expression for the free energy $f(k)$ of one more term (in comparison with Eq. (12.49)):

$$f(k) = -\frac{1}{2}\beta k V - \sqrt{\frac{V}{2\beta k}}\,(1-k) + \frac{1}{6}(\beta k)^4 V^2 r^2 \tag{12.56}$$

which describes the contribution related to mutual interaction between the blocks. It can be easily checked that under condition $r^2 \ll 1$ the solution of the corresponding saddle-point equation (obtained by variation of Eq. (12.56) with respect to k)

$$-V + \sqrt{\frac{V}{2(\beta k)^3}}\,(1+k) + \frac{4}{3}(\beta k)^3 V^2 r^2 = 0 \tag{12.57}$$

is still given by Eq. (12.51):

$$\beta k_* \approx (2V)^{-1/3} + O(r^2) \tag{12.58}$$

The applicability of this approach requires that the distances between the 'blocks' (consisting of k particles) should be much larger than the size of the well r (exactly in the same way as when a Kardar-type solution is constructed from the separate particles and not from the blocks). In the present case the solution of the corresponding Kardar problems for n/k block particles is described by the wave function (12.26) where the factor $\beta^3 u$ should be substituted by $(\beta k)^3 Vr$. The condition that the typical separation between these block particles must be much greater than the size of the well: $(\beta k)^{-3}(Vr)^{-1} \gg r$, for the saddle-point value of k, Eq. (12.58), yields: $r^2 \ll 1$. This observation confirms that the solution obtained above is selfconsistent.

The form of Eq. (12.55) shows that in the case considered the only non-linear (in n) contribution to $E(n)$ is also of the third order in n. In terms of the general representation (12.15) this corresponds to

$$\beta E_3 = -V^2 r^2 (\beta k)^2 \beta^3 \tag{12.59}$$

Using the saddle-point value of the parameter k, (Eq. (12.58)), one gets:

$$\beta E_3(k_*) \simeq -\frac{1}{4}(2V)^{4/3} r^2 \beta^3 \tag{12.60}$$

Correspondingly, according to Eq. (12.17), for the typical value of the free energy fluctuations one obtains the following temperature-independent result:

$$\delta F \sim V^{4/9} r^{2/3} L^{1/3} \tag{12.61}$$

Finally, for the mean square deviation of the polymer trajectory $\overline{\langle x^2 \rangle} \sim \delta F L$ one finds the scaling law which corresponds to the wandering exponent $\zeta = 2/3$:

$$\overline{\langle \phi^2 \rangle} \sim V^{4/9} r^{2/3} L^{4/3} \tag{12.62}$$

Note that unlike the corresponding high-temperature solution, Eq. (12.42), the prefactor of the above scaling law appears to be temperature independent.

Thus we have demonstrated that in the case when random potential correlations are characterized by a finite correlation radius r, the solution at low temperatures has an essentially different structure than at high-temperatures. Nonetheless the value of the wandering exponent in both cases is the same: $\zeta = 2/3$. In contrast to the high-temperature limit for which the prefactor in the scaling law $\overline{\langle x^2 \rangle} \sim aL^{2\zeta}$ is temperature dependent: $a \propto T^{-2/3}$, in the low-temperature limit it saturates at finite value $a \propto r^{2/3}$.

Because the value of the wandering exponent for both regimes is the same, there are no reasons for the sharp transition between these two types of solution. At very low temperatures the solution is characterized by the one-step replica symmetry breaking; that is, the replicas are split into n/k well separated blocks. As the temperature grows, the distance between the blocks becomes comparable with the size of each block. At higher temperatures the replica symmetry breaking phenomenon can manifest itself only in a slight modulation of the distance between nearest replicas in comparison with what follows from the 'replica symmetric' wave function (12.26).

Note in conclusion, that in this chapter we have considered only the simplest type of the directed polymers problems, for which the results for the wandering exponent are believed to be sufficiently well established. Nevertheless, even here, in the one-dimensional case with locally correlated random potential, there are important problems that are still waiting to be solved. Presumably the most important one is that the Bethe ansatz type of solution considered in this chapter (as well as the equivalent type of solutions in terms of Burgers' equation [85]) does not allow us to calculate the scaling of $\overline{\langle\phi^2(L)\rangle}$ in a *direct way*, as represented in Eq. (12.3). Moreover, one can easily check that this type of solution can not provide any reasonable result even for the replica partition function of the originally formulated problem with free boundary conditions at $x = L$, Eq. (12.19): the integral over $\{y_a\}$ of the wave function (12.26) is divergent! Of course, this divergency can be cured by taking into account the degree of freedom connected to the center of mass of the particles, which in the considered solution was supposed to be zero. But then, one finds that replica partition function (12.26) appears to be proportional to $\sqrt{n}$, and therefore it has no reasonable $n \to 0$ limit! Actually all these troubles are the consequence of the crucial assumption that the leading contribution to the 'wave function', the partition function $Z_L(\mathbf{y})$, Eq. (12.20), comes only from the ground-state solution of the quantum-mechanical problem (12.21). This assumption must indeed be correct until the coordinates of the end-points $\{y_a\}$ are much smaller than L (in other words the limit $L \to \infty$ must be taken first while the values of $\{y_a\}$ are kept finite). In fact, in the calculation of the total replica partition function (12.20) (as well as in the direct calculation of $\overline{\langle\phi^2(L)\rangle}$) one has to integrate over $\{y_a\}$, which scales as a power of L. Therefore in the correct selfconsistent solution of this problem one has to take into account the whole *spectrum* of excited states hidden in the Shrödinger equation (12.21). At the moment this *selfconsistent* solution is not known.

Even less clear is the situation in more sophisticated types of directed polymer model. In particular, there are no solutions for the one-dimensional systems with the random potential characterized by (slow) long-distance algebraic decay of correlations, not to mention models in higher dimensions.

13
Vector breaking of replica symmetry

In this chapter we present a new method for studying statistical systems with quenched disorder in the low-temperature limit [88]. The use of the replica method has turned out to be very efficient in some disordered systems. It allows for a detailed characterization of the low-temperature phase at least at the mean-field level. In all the mean-field spin-glass-like problems where one can expect the mean-field theory to be exact, the Parisi scheme of replica symmetry breaking [1] is successful, and at the moment there is no counterexample showing that it does not work. On the other hand, the low-temperature phase of these systems is complicated enough, even at the mean-field level. One might hope that the very low-temperature limit could be easier to analyse, while its physical content should be basically the same. This very low-temperature limit is also an extreme case where one might hope to get a better understanding of the finite-dimensional problems. At first sight the low-temperature limit is indeed simpler because the partition function could be analysed at the level of a saddle-point approximation. However, it is easy to see that generically this limit does not commute with the limit of the number of replicas going to zero. There is a very basic origin to this non-commutation, namely the fact that there still exist, even at zero temperature, sample-to-sample fluctuations.

In this chapter we develop a new method of summation over all saddle points of a corresponding replica Hamiltonian, in order to investigate the low-temperature behavior of glassy systems. It appears that in a large class of strongly disordered systems, it is necessary to include saddle points of the Hamiltonian that break the replica symmetry in a *vector* sector, as opposed to the usual matrix sector breaking of spin-glass mean-field theory. First, we shall demonstrate how this new method works on the examples of some elementary (zero-dimensional) problems, which can be solved directly. As for its application to more difficult problems, we obtain very good approximation to the zero temperature fluctuations of a particle in a random potential, and we also discuss the case of directed polymers in random media with long-range correlations, where we rederive the scaling exponents using this new method.

Additionally, in the last section of this chapter we consider the problem of the existence of the non-analytic (Griffiths-like) contributions to the free energy of a weakly disordered Ising ferromagnet (in the zero external magnetic field) from the point of view of the replica theory. Here it is demonstrated that in the paramagnetic phase (away from the critical point) such contributions appear as a result of non-linear localized (instanton-like) solutions of the mean-field stationary equations, which are characterized by the vector type of the replica symmetry breaking. It is argued that physically these replica instantons describe the contribution from rare spatial 'ferromagnetic islands' in which the local (random) temperature is below T_c.

13.1 Zero-dimensional systems

13.1.1 The Ising model

To demonstrate in the simplest terms how the proposed procedure works, we consider first some trivial zero-dimensional problems. The simplest example is one Ising spin $\sigma = \pm 1$ in a random field h. The Hamiltonian is:

$$H = \sigma h \tag{13.1}$$

where the distribution of the random field is Gaussian:

$$P(h) = \frac{1}{\sqrt{2\pi h_0^2}} \exp\left(-\frac{h^2}{2h_0^2}\right) \tag{13.2}$$

The free energy is:

$$-\beta F(h_0;\beta) = \overline{\ln\left[\sum_{\sigma=\pm 1} \exp(-\beta\sigma h)\right]} = \int_{-\infty}^{+\infty} Dx \ln[1+\exp(2\beta h_0 x)] \tag{13.3}$$

where Dx is the centred Gaussian measure of width one:$Dx \equiv dx/\sqrt{2\pi}\exp(-\frac{1}{2}x^2)$. In particular, in the zero temperature limit one finds:

$$F(h_0;\beta\to\infty) = -\frac{2h_0}{\sqrt{2\pi}} \tag{13.4}$$

Let us consider now how this 'problem' can be solved in terms of the replica approach:

$$-\beta F(h_0;\beta) = \lim_{n\to 0} \frac{1}{n}(\overline{Z^n} - 1)$$

$$\lim_{n\to 0} \frac{1}{n}\left[\sum_{\{\sigma_a\}=\pm 1} \exp\{\frac{1}{2}\beta^2 h_0^2 (\sum_{a=1}^{n} \sigma_a)^2\} - 1\right]$$

$$\lim_{n\to 0} \frac{1}{n}\left[\sum_{k=0}^{n} \frac{n!}{k!(n-k)!} \exp\{\frac{1}{2}\beta^2 h_0^2 (2k-n)^2\} - 1\right] \quad (13.5)$$

In view of the application of the method to more complicated problems we want to compute the behavior at low temperature. This cannot be done naively from a saddle-point evaluation of the sum at large β, because of the non-commutativity of the limits $\beta \to \infty$ and $n \to 0$. Instead we proceed as follows. The term $k = 0$, which is the contribution from the 'replica symmetric (RS) configuration' $\sigma_a = +1$, is singled out; its contribution is equal to $1 + O(n^2)$, which cancels the (-1) in Eq. (13.5). The contributions of the rest of the terms (which could be interpreted as corresponding to the states with 'replica symmetry breaking' (RSB) in the replica vector $\{\sigma_a\}$) can be represented as follows:

$$-\beta F(h_0;\beta) = \lim_{n\to 0} \frac{1}{n} \sum_{k=1}^{\infty} \frac{\Gamma(n+1)}{\Gamma(k+1)\Gamma(n-k+1)} \exp\left[\frac{1}{2}\beta^2 h_0^2 (2k-n)^2\right] \quad (13.6)$$

Here the summation over k can be extended beyond $k = n$ to ∞, because the gamma function is equal to infinity at negative integers. Now using the relation $\Gamma(-z) = \pi[z\Gamma(z)\sin(\pi z)]^{-1}$, we can perform the analytic continuation $n \to 0$:

$$\frac{\Gamma(n+1)}{\Gamma(k+1)\Gamma(n-k+1)} = \frac{\Gamma(n+1)(k-1-n)\Gamma(k-1-n)\sin(\pi(k-1-n))}{\pi\Gamma(k+1)}\Big|_{n\to 0}$$

$$\simeq n\frac{(-1)^{k-1}}{k} \quad (13.7)$$

Thus, for the free energy (13.6) one obtains:

$$-\beta F(h_0;\beta) = \sum_{k=1}^{\infty} \frac{(-1)^{k-1}}{k} \exp\left(2\beta^2 h_0^2 k^2\right) = \int_{-\infty}^{+\infty} Dx \ln\left[1 + \exp(2\beta h_0 x)\right] \quad (13.8)$$

We see that this result coincides with the one (13.3) obtained by the direct calculation. This is of course no surprise since we have just done an exact replica computation. But it exemplifies some of the steps that we shall need below, in particular the proper definition and computation of the divergent series appearing in (13.8) through an integral representation.

13.1.2 The 'soft' Ising model

Consider now the 'soft' version of the Ising model described by the double-well Hamiltonian:

$$H = -\frac{1}{2}\tau\phi^2 + \frac{1}{4}\phi^4 - h\phi \tag{13.9}$$

where the random field is described by the Gaussian distribution (13.2). We concentrate again on the zero temperature limit. Besides, we assume that the typical value of the field h_0 is small ($h_0 \ll \tau^{3/2}$). In this case the field will not destroy the double-well structure of the Hamiltonian (13.9), and (at $T \to 0$) the system must be equivalent to the discrete Ising model considered before. (The 'opposite limit' of the random field Hamiltonian with only one ground state will be considered in Section 13.1.3).

The direct calculation of the zero-temperature free energy is trivial. For a given value of the field $h \ll \tau^{3/2}$ the ground states of the Hamiltonian (13.9) are: $\phi_1 \simeq +\sqrt{\tau} + h/2\tau$, for $h > 0$; and $\phi_1 \simeq -\sqrt{\tau} + h/2\tau$, for $h < 0$. In both cases the corresponding energy is $E_g(h) \simeq -\frac{1}{4}\tau^2 - |h|\sqrt{\tau}$. Thus, the zero-temperature averaged free energy is:

$$F(h_0) \simeq -\frac{1}{4}\tau^2 - 2\sqrt{\tau}\int_0^{+\infty} \frac{dh}{\sqrt{2\pi h_0^2}} h \exp\left(-\frac{h^2}{2h_0}\right) = -\frac{1}{4}\tau^2 - \frac{2h_0\sqrt{\tau}}{\sqrt{2\pi}} \tag{13.10}$$

Consider now how this result can be obtained in terms of replicas. The replica Hamiltonian and the corresponding saddle-point equations are:

$$H_n = -\frac{1}{2}\tau\sum_{a=1}^{n}\phi_a^2 + \frac{1}{4}\sum_{a=1}^{n}\phi_a^4 - \frac{1}{2}\beta h_0^2\left(\sum_{a=1}^{n}\phi_a\right)^2 \tag{13.11}$$

$$-\tau\phi_a + \phi_a^3 = \beta h_0^2\left(\sum_{a=1}^{n}\phi_a\right) \tag{13.12}$$

The 'replica-symmetric' solution of these equations (in the limit $n \to 0$) is: $\phi_a = \phi_{rs} = \sqrt{\tau}$. The corresponding energy is $E_{rs} = -\frac{1}{4}n\tau^2$. This solution (in the limit $n \to 0$) does not involve the contribution from the random field.

Proceeding along the lines of Section 13.1.1, we have to look also for the solutions of Eqs. (13.12) which would involve the 'replica symmetry breaking' in the replica vector $\{\phi_a\}$:

$$\phi_a = \begin{cases} \phi_1 & \text{for } a = 1, \dots, k \\ \phi_2 & \text{for } a = k+1, \dots, n \end{cases} \tag{13.13}$$

In terms of this ansatz in the limit $n \to 0$ the replica summations can be performed according to the following simple rule: $\sum_{a=1}^{n}\phi_a = k\phi_1 +$

$(n-k)\phi_2 \to k(\phi_1 - \phi_2)$. The saddle-point Eqs. (13.12) then turn into two equations for ϕ_1 and ϕ_2:

$$-\tau\phi_{1,2} + \phi_{1,2}^3 = \beta k h_0^2(\phi_1 - \phi_2) \tag{13.14}$$

Assuming that $\beta k h_0^2 \ll \tau$ (the explanation of this strange assumption – considering that we are interested in the $\beta \to \infty$ limit! – will be given below), in the leading order one gets:

$$\phi_1 \simeq +\sqrt{\tau}; \quad \phi_2 \simeq -\sqrt{\tau} \tag{13.15}$$

From Eq. (13.11) one obtains the corresponding energy of the above 'RSB' saddle-point solution:

$$\begin{aligned} E_k &= -\frac{1}{2}k\tau(\phi_1^2 - \phi_2^2) + \frac{1}{4}k(\phi_1^4 - \phi_2^4) - \frac{1}{2}\beta h_0^2 k^2(\phi_1 - \phi_2)^2 \\ &\simeq -2\beta k^2 h_0^2 \tau \ + \ O(h_0^4) \end{aligned} \tag{13.16}$$

Now, summing the contributions from all these saddle points similarly to the calculations of Section 13.1.1 for the zero-temperature free energy one obtains:

$$\begin{aligned} F(h_0) &= -\lim_{n\to 0}\frac{1}{\beta n}(Z_n - 1) \\ &\simeq -\lim_{n\to 0}\frac{1}{\beta n}(Z_{\text{'rs'}} - 1) - \lim_{n\to 0}\frac{1}{\beta n}Z_{\text{'rsb'}} \\ &= -\lim_{n\to 0}\frac{1}{\beta n}\left[\exp(\frac{1}{4}\beta n\tau^2) - 1\right] - \frac{1}{\beta}\sum_{k=1}^{\infty}\frac{(-1)^{k-1}}{k}\exp\left(2\beta^2 k^2 h_0^2\tau\right) \\ &= -\frac{1}{4}\tau^2 - \frac{1}{\beta}\int_{-\infty}^{+\infty}\frac{dx}{\sqrt{2\pi}}\exp(-\frac{1}{2}x^2)\ln\left[1 + \exp\left(2\beta h_0 x\sqrt{\tau}\right)\right] \end{aligned} \tag{13.17}$$

Taking the limit $\beta \to \infty$ one finally gets the result:

$$F(h_0) \simeq -\frac{1}{4}\tau^2 - \frac{1}{\beta}2\beta h_0\sqrt{\tau}\int_0^{+\infty}\frac{dx}{\sqrt{2\pi}}x\exp\left(-\frac{1}{2}x^2\right) = -\frac{1}{4}\tau^2 - \frac{2h_0\sqrt{\tau}}{\sqrt{2\pi}} \tag{13.18}$$

which coincides with Eq. (13.10).

It is worth noting that the summation of the series in Eq. (13.17) can also be performed another way:

$$F_{\text{rsb}} = -\frac{1}{\beta}\sum_{k=1}^{\infty}\frac{(-1)^{k-1}}{k}\exp\left(2\beta^2 k^2 h_0^2\tau\right) = \frac{1}{2i\beta}\int_C\frac{dz}{z\sin(\pi z)}\exp\left(2\beta^2 z^2 h_0^2\tau\right) \tag{13.19}$$

where the integration goes over the contour in the complex plane shown in Fig. 13.1(a). Then we can move the contour to the position shown in Fig. 13.1(b), and after the change of integration variable:

$$z \to \left[2\beta^2 h_0^2 \tau\right]^{-1/2} ix \tag{13.20}$$

in the limit $\beta \to \infty$ we have:

$$\sin(\pi z) \simeq \frac{i\pi x}{\beta\sqrt{2h_0^2\tau}} \tag{13.21}$$

Then, taking into account also the contribution from the pole at $x = 0$ for the integral in Eq. (13.19) we get:

$$F_{\text{rsb}} = \frac{h_0\sqrt{2\tau}}{2\pi}\int_{-\infty}^{+\infty}\frac{dx}{x^2}\left[\exp(-x^2) - 1\right] = -\frac{2h_0\sqrt{\tau}}{\sqrt{2\pi}} \tag{13.22}$$

which again coincides with the results (13.10) and (13.18).

This little exercise with the integral representation of the divergent series in Eq. (13.19) shows in particular that the 'effective' value of the parameter $\beta k \to \beta z$ which enters into the saddle-point equations (13.14) scales (according to (13.20)) as $(h_0\sqrt{\tau})^{-1}$. That is why, in the zero-temperature limit, the 'effective' value of the factor $\beta k h_0^2 \sim h_0/\sqrt{\tau}$ in Eq. (13.14) can be assumed to be small compared with τ (for small fields $h_0 \ll \tau^{3/2}$).

Replica fluctuations

Because of the non-commutativity of the limits $n \to 0$ and $\beta \to \infty$, one cannot get the exact result by keeping only the saddle-point states of the replica Hamiltonian. Actually, averaging over quenched disorder involves the effect of sample-to-sample fluctuations, which in terms of the replica formalism manifest itself as the contribution from the replica fluctuations. In other words, to get an exact result in terms of replicas the contribution from the saddle points is not enough, and one has to integrate over replica fluctuations even in the zero-temperature limit.

This phenomenon can be easily demonstrated for the above example of the 'soft' Ising model. Let us take into account the contribution from the Gaussian replica fluctuations near the 'replica-symmetric' saddle point $\phi_a = \phi_{\text{rs}} = \sqrt{\tau}$:

$$\phi_a = \phi_{\text{rs}} + \varphi_a \tag{13.23}$$

From Eq. (13.11) for the 'replica-symmetric' part of the partition function

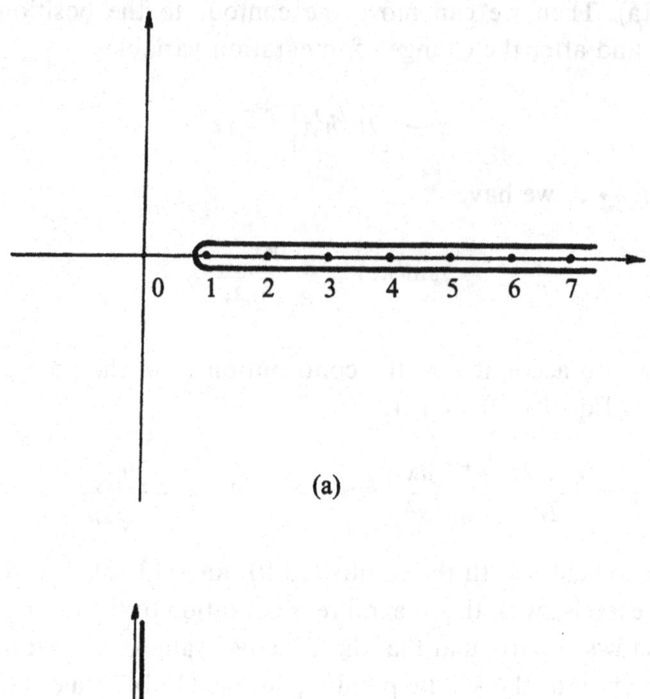

(a)

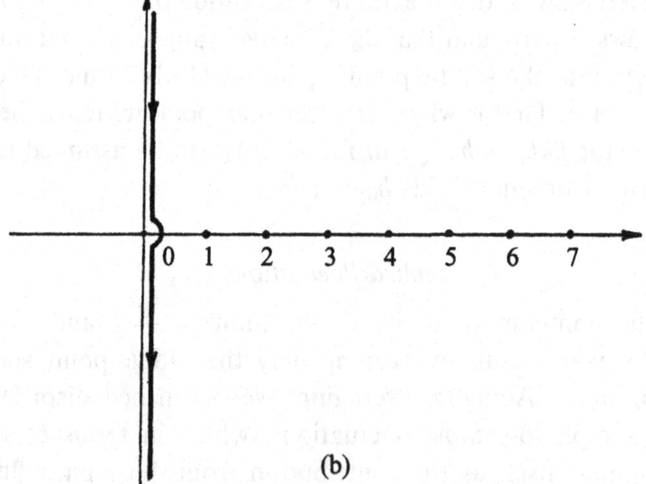

(b)

Fig. 13.1. The contours of integration in the complex plane used for summing the series: (a) the original contour; (b) the deformed contour.

we get:

$$Z_{\text{'rs'}} = \exp\left(\frac{1}{4}\beta n\tau^2\right) \int D\varphi_a \exp\left[-\beta \sum_{a,b}^{n} (\tau\delta_{ab} - \frac{1}{2}\beta h_0^2)\varphi_a\varphi_b\right]$$

$$\simeq \exp\left[\frac{1}{4}\beta n\tau^2 + \frac{\beta n h_0^2}{4\tau} - \frac{1}{2} n\ln(\beta\tau)\right] \qquad (13.24)$$

Therefore, in the zero-temperature limit one obtains the following contribution to the free energy:

$$F_{\text{`rs'}} = -\frac{1}{4}\tau^2 - \frac{h_0^2}{4\tau} + O(h_0^4) \tag{13.25}$$

After some more work one can also derive the contribution from replica fluctuations near the RSB solution (13.13) (we omit here these couple of pages of slightly painful algebra):

$$\delta F_{\text{`rsb'}} = +\frac{h_0^3}{2\sqrt{2\pi}\,\tau^{5/2}} + O(h_0^5) \tag{13.26}$$

Thus, in order h_0^3 the zero-temperature free energy of this system is:

$$F(h_0) = -\frac{1}{4}\tau^2 - \frac{2h_0\sqrt{\tau}}{\sqrt{2\pi}} - \frac{h_0^2}{4\tau} + \frac{h_0^3}{2\sqrt{2\pi}\,\tau^{5/2}} + O(h_0^4) \tag{13.27}$$

This result can be easily confirmed by direct calculation of the averaged ground-state energy of the Hamiltonian (13.9).

We see that at $T = 0$ there exist finite contributions $\sim h_0^2/\tau$ and $\sim h_0^3/\tau^{5/2}$ due to the replica fluctuations. In this particular example the value of h_0 was assumed to be small, and these contributions can be treated as a small correction. However, we should keep in mind that the contribution from the replica fluctuations in general could appear to be of the same order as that from the saddle points. Therefore, the calculations we are going to perform in the next sections for less trivial examples, taking into account only saddle-point states, can not pretend to give exact results, giving only the scaling dependence from the parameters of a model.

Saddle points

In the above calculations of the free energy for the 'soft' Ising system we have taken into account only the contribution from the two *minima* of the double-well potential. The existence of the third saddle point, which is the maximum at $\phi = 0$, has been ignored. In this particular example such an algorithm looks natural. However, in less trivial systems very often it is not easy to distinguish the types of saddle point involved. Moreover, it could be very hard to impose a simple and robust 'discrimination' rule with respect to different types of saddle point, which would not block the calculations at the very start.

Because of this, we would like to propose a somewhat modified scheme of calculations, which takes into account *all* saddle points. In the above example of the 'soft' Ising model the third saddle point (the maximum) is at

$\phi = 0$. Then, instead of the ansatz (13.13) let us represent the replica vector ϕ_a as follows:

$$\phi_a = \begin{cases} +\sqrt{\tau} & \text{for } a = 1,\ldots,k \\ -\sqrt{\tau} & \text{for } a = k+1,\ldots,k+l \\ 0 & \text{for } a = k+l+1,\ldots,n \end{cases} \tag{13.28}$$

The corresponding energy (in the limit $n \to 0$) is:

$$H_{kl} = -\frac{1}{4}\tau^2(k+l) - \frac{1}{2}\beta h_0^2 \tau(k-l)^2 + O(h_0^4) \tag{13.29}$$

Note that in terms of the ansatz (13.28) the 'replica symmetric' state ($k = l = 0$), $\phi_a = \phi_0 = 0$ has zero energy, so that it gives no contribution to the free energy.

The combinatoric factor in the $n \to 0$ limit is now:

$$\frac{n!}{k!\,l!(n-k-l)!} \to n\frac{(-1)^{k+l-1}}{k+l}\frac{(k+l)!}{k!\,l!} \tag{13.30}$$

Thus, for the free energy (for $\beta \to \infty$) we obtain:

$$F(h_0) = -\frac{1}{\beta}\sum_{k+l=1}^{\infty}\frac{(-1)^{k+l-1}}{k+l}\frac{(k+l)!}{k!l!}\exp\left[\frac{1}{4}\beta\tau^2(k+l) + \frac{1}{2}\beta^2 h_0^2\tau(k-l)^2\right] \tag{13.31}$$

This series can be summed up in a similar way as those in Eqs. (13.8) and (13.17):

$$F(h_0) = -\frac{1}{\beta}\int_{-\infty}^{+\infty} Dx \ln\left[1 + \exp\left(\frac{1}{4}\beta\tau^2 + \beta h_0\sqrt{\tau}x\right) + \exp\left(\frac{1}{4}\beta\tau^2 - \beta h_0\sqrt{\tau}x\right)\right] \tag{13.32}$$

In the limit $\beta \to \infty$ one finds:

$$F(h_0) = -\frac{1}{\beta}\int_{-\infty}^{+\infty} Dx\left[\frac{1}{4}\beta\tau^2 + \beta h_0\sqrt{\tau}|x|\right] = -\frac{1}{4}\tau^2 - \frac{2h_0\sqrt{\tau}}{\sqrt{2\pi}} \tag{13.33}$$

Again, we get the correct result. Although it might seem at first sight somewhat 'magic', at least some aspects of this computation can be understood. In the example considered (as well as in the further examples to be studied below) the relevant states, which contribute to the free energy, have negative energy $-E(h)$. Then, in the low-temperature limit the partition function of a given sample is $Z \simeq \exp(+\beta E(h))$. Therefore, in the limit $\beta \to \infty$ the free energy can be represented with exponential accuracy as follows:

$$F(h_0) = -\frac{1}{\beta}\ln Z = -\frac{1}{\beta}\ln\left(\frac{1+Z}{1+Z^{-1}}\right)$$

$$= -\frac{1}{\beta}\overline{\ln(1+Z)} + \frac{1}{\beta}\overline{\ln(1+Z^{-1})}$$
$$= -\frac{1}{\beta}\sum_{m=1}^{\infty}\frac{(-1)^{m-1}}{m}\overline{Z^m} + \frac{1}{\beta}\sum_{m=1}^{\infty}\frac{(-1)^{m-1}}{m}\overline{Z^{-m}}$$
$$\simeq -\frac{1}{\beta}\sum_{m=1}^{\infty}\frac{(-1)^{m-1}}{m}\overline{Z^m} \tag{13.34}$$

One can easily check that after averaging $\overline{Z^m} \equiv Z_m$ and taking into account the contributions of the two minima of the corresponding replica Hamiltonian H_m one recovers the series in Eq. (13.31).

The only 'magic' rule that should be followed in the direct replica calculations is that the 'background' state, $\phi_0 = 0$ (the one with zero energy) in the ansatz for the replica vector ϕ_a of the type (13.28) should be placed in the *last* group of replicas. Using this rule, the series obtained for the free energy will correspond to the above interpretation (13.34).

13.1.3 One-well potential

Consider now how the method works in the case where the Hamiltonian has only one minimum:

$$H = \frac{1}{\alpha}\phi^{\alpha} - h\phi \tag{13.35}$$

where $\phi \geq 0$ and $\alpha \geq 2$, and the random field h is again described by the Gaussian distribution (13.2). For $\alpha = 4$ this system can be interpreted as the variant of the Hamiltonian (13.9) in the limit of strong magnetic fields.

In the zero-temperature limit the free energy is defined by the ground state $\phi(h) = h^{1/(\alpha-1)}$ for $h > 0$, and $\phi = 0$ for $h \leq 0$. Its energy is $E(h) = -[(\alpha-1)/\alpha]h^{\alpha/(\alpha-1)}$ for $h > 0$, and $E = 0$ for $h \leq 0$. Therefore, for the averaged zero-temperature free energy we find:

$$F(h_0) = -\frac{\alpha-1}{\alpha}\int_0^{+\infty}\frac{dh}{\sqrt{2\pi h_0^2}}h^{\frac{\alpha}{\alpha-1}}\exp\{-\frac{h^2}{2h_0^2}\}$$
$$= -\left[\frac{\alpha-1}{\alpha}\int_0^{+\infty}\frac{dx}{\sqrt{2\pi}}x^{\frac{\alpha}{\alpha-1}}\exp\left(-\frac{1}{2}x^2\right)\right]h_0^{\frac{\alpha}{\alpha-1}}$$
$$\equiv -A(\alpha)\,h_0^{\frac{\alpha}{\alpha-1}} \tag{13.36}$$

In terms of replicas, the corresponding replicated Hamiltonian is:

$$H_n = \frac{1}{\alpha}\sum_{a=1}^{n}\phi_a^{\alpha} - \frac{1}{2}\beta h_0^2\sum_{a,b=1}^{n}\phi_a\phi_b \tag{13.37}$$

This Hamiltonian has a trivial 'background' extremum at $\phi = 0$ with zero energy. Therefore, following the scheme proposed in the previous subsection, we look for non-trivial saddle-point solutions in terms of the following ansatz:

$$\phi_a = \begin{cases} \phi & \text{for } a = 1,\ldots,k \\ 0 & \text{for } a = k+1,\ldots,n \end{cases} \tag{13.38}$$

For the corresponding Hamiltonian and the saddle-point equation (in the limit $n \to 0$) one gets:

$$H_k = \frac{1}{\alpha}k\phi^\alpha - \frac{1}{2}\beta h_0^2 k^2 \phi^2 \tag{13.39}$$

$$\phi^{\alpha-1} - \beta h_0^2 k \phi = 0 \tag{13.40}$$

The solution to this equation and the corresponding energy are:

$$\phi = (\beta k h_0^2)^{\frac{1}{\alpha-2}} \tag{13.41}$$

$$H_k = -\frac{1}{\beta}\frac{\alpha-2}{2\alpha}(\beta k)^{\frac{2(\alpha-1)}{\alpha-2}} h_0^{\frac{2\alpha}{\alpha-2}} \tag{13.42}$$

(Note that although one can try with more 'RSB' steps in the replica vector ϕ_a it can be easily proved that there exists only one type of the non-trivial solution given by the ansatz (13.38).) Then, in terms of the procedure described above for the free energy we have:

$$F(h_0) = -\frac{1}{\beta}\sum_{k=1}^{\infty}\frac{(-1)^{k-1}}{k}\exp\left[\frac{\alpha-2}{2\alpha}(\beta k)^{\frac{2(\alpha-1)}{\alpha-2}} h_0^{\frac{2\alpha}{\alpha-2}}\right] \tag{13.43}$$

The summation of this series can be performed in terms of the integral representation Eq. (13.19):

$$F(h_0) = \frac{1}{2i\beta}\int_C \frac{dz}{z\sin(\pi z)}\exp\left[\frac{\alpha-2}{2\alpha}(\beta z)^{\frac{2(\alpha-1)}{\alpha-2}} h_0^{\frac{2\alpha}{\alpha-2}}\right] \tag{13.44}$$

where the integration goes over the contour in the complex plane shown in Fig. 13.1(a). Then, again, we move the contour to the position shown in Fig. 13.1(b) and redefine the integration variable:

$$z \to \frac{1}{\beta}h_0^{-\frac{\alpha}{\alpha-1}} ix \tag{13.45}$$

In the limit $\beta \to \infty$ we have:

$$\sin(\pi z) \simeq \frac{i\pi x}{\beta}\, h_0^{-\frac{\alpha}{\alpha-1}} \tag{13.46}$$

and

$$F(h_0) = -\left[\frac{1}{2\pi}\int_{C_1}\frac{dx}{x^2}\exp\left(\frac{\alpha-2}{2\alpha}(ix)^{\frac{2(\alpha-1)}{\alpha-2}}\right)\right] h_0^{\frac{\alpha}{\alpha-1}} \equiv -B(\alpha)\, h_0^{\frac{\alpha}{\alpha-1}} \tag{13.47}$$

Thus, we have obtained the correct scaling of the free energy as a function of h_0. Note, however, that although it is also possible to calculate the value of the coefficient $B(\alpha)$ in the integral (13.47), such a calculation would not make much sense because to obtain the correct coefficient (which is given by the integral in (13.36)) one would need to take into account replica fluctuations, which we have neglected here.

13.1.4 The toy model

Let us consider now a slightly less trivial example of a zero-dimensional system, which cannot be solved by elementary algebra. This system, generally called the 'toy model', consists of a single degree of freedom ϕ evolving in an energy landscape that is the sum of a quadratic well and a Brownian potential. The Hamiltonian is:

$$H = \frac{1}{2}\mu\phi^2 + V(\phi) \tag{13.48}$$

where $V(\phi)$ is the random potential described by the Gaussian distribution:

$$P[V(\phi)] \sim \exp\left[-\frac{1}{4g}\int d\phi\left(\frac{dV}{d\phi}\right)^2\right] \tag{13.49}$$

The V distribution is characterized by its first two moments:

$$\begin{aligned}\overline{(V(\phi)-V(\phi'))^2} &= 2g|\phi-\phi'|;\\ \overline{V(\phi)} &= 0;\\ \overline{V(\phi)V(\phi')} &= C - g|\phi-\phi'|\end{aligned} \tag{13.50}$$

where C is an irrelevant constant. This problem was introduced originally as a toy, zero-dimensional version of the interface in the random field Ising model [89]. It has the virtue of showing explicitly how the most standard field theoretic methods like perturbation theory, iteration methods or supersymmetry get fooled in this problem, in the low-μ, low-temperature limit, by the existence of many metastable states [90–93]. The main point is that at low enough temperatures the usual perturbation theory does not work and a qualitatively reasonable theory must involve the effects of the replica symmetry breaking. This has been demonstrated within the replica Gaussian variational approximation [83, 94].

One quantity that one would like to calculate in such a system is the value of $\overline{\langle\phi^2\rangle}$ in the limit of the zero temperature. Using simple energy arguments one can easily estimate what must be the scaling dependence of this quantity on the parameters μ and g. For a given value of ϕ the loss of energy due to the attractive potential in the Hamiltonian (13.48) is $\sim \mu\phi^2$. The possible gain of energy due to the random potential according to statistics (13.49)–(13.50) can be estimated as $\sim \sqrt{g}\sqrt{\phi}$. Optimizing the total energy $E \sim \mu\phi^2 - \sqrt{g}\sqrt{\phi}$ with respect to ϕ one finds that

$$\overline{\langle\phi^2\rangle} = C_2 \frac{g^{2/3}}{\mu^{4/3}} \tag{13.51}$$

This result tells us that the typical energy minimum of the Hamiltonian (13.48) lies at a finite distance from the origin. The scaling (13.51), which is obviously correct, is not so easy to derive from some field theoretic methods which could be also used in higher dimensions, and there is no known exact result for the constant C_2 at the moment.

Let us try to calculate the value of $\overline{\langle\phi^2\rangle}$ in the zero temperature limit using the method considered above. The replicated Hamiltonian of the system (13.48) is:

$$H_n = \frac{1}{2}\mu\sum_{a=1}^{n}\phi_a^2 + \frac{1}{2}\beta g \sum_{a,b=1}^{n} |\phi_a - \phi_b| \tag{13.52}$$

The corresponding saddle-point equations are:

$$\mu\phi_a + \beta g \sum_{b=1}^{n} \mathrm{Sign}\,(\phi_a - \phi_b) = 0 \tag{13.53}$$

(Note that in this formula, whenever there is some ambiguity, one should always assume that there is at some intermediate step a short-scale regularization. Therefore, one must interpret for instance Sign(0) = 0.) Let us first look for non-trivial solutions of Eqs. (13.53). It can be easily proven that within the 'one-step' RSB ansatz (13.13) there exist no non-trivial solutions. Let us consider the 'two-steps' ansatz for the replica vector ϕ_a:

$$\phi_a = \begin{cases} \phi_1 & \text{for } a = 1,\ldots,k \\ \phi_2 & \text{for } a = k+1,\ldots,k+l \\ \phi_3 & \text{for } a = k+l+1,\ldots,n \end{cases} \tag{13.54}$$

From Eqs. (13.53) one finds the following equations for $\phi_{1,2,3}$ (in the limit $n \to 0$):

$$\mu\phi_1 + \beta g l \mathrm{Sign}(\phi_1 - \phi_2) - \beta g(k+l)\mathrm{Sign}(\phi_1 - \phi_3) = 0$$
$$\mu\phi_2 + \beta g k \mathrm{Sign}(\phi_2 - \phi_1) - \beta g(k+l)\mathrm{Sign}(\phi_2 - \phi_3) = 0$$

$$\mu\phi_3 + \beta gk\text{Sign}(\phi_3 - \phi_1) + \beta gl\text{Sign}(\phi_3 - \phi_2) = 0 \tag{13.55}$$

The solution of these equations is:

$$\phi_1 = -\frac{g}{\mu}\beta k; \quad \phi_2 = +\frac{g}{\mu}\beta l; \quad \phi_3 = \frac{g}{\mu}\beta(l-k) \tag{13.56}$$

and the corresponding energy is (in the limit $n \to 0$):

$$E_{kl} = -\frac{\beta^2 g^2}{2\mu} kl(k+l) \tag{13.57}$$

It can be proven that there exist no other solutions of the saddle-point equations (13.53) with a number of RSB steps larger than two. Therefore, (after taking the limit $n \to 0$) for the 'RSB' part of the free energy we get the following series (see Eq. (13.30)):

$$\begin{aligned} F_{\text{rsb}} &= -\frac{1}{\beta n}\sum_{k+l=1}^{n} \frac{n!}{k!l!(n-k-l)!}\exp(-\beta E_{kl}) \\ &\to -\frac{1}{\beta}\sum_{k+l=1}^{\infty} \frac{(-1)^{k+l-1}}{k+l}\frac{(k+l)!}{k!l!}\exp\left[\lambda kl(k+l)\right] \end{aligned} \tag{13.58}$$

where

$$\lambda = \frac{\beta^3 g^2}{2\mu} \to \infty \tag{13.59}$$

We again carry the summation of the asymptotic series (13.58) with the integral method mentioned in Section 13.1.2:

$$\begin{aligned} F_{\text{rsb}} &= \frac{1}{\beta(2i)^2}\int\int_C \\ &\times \frac{dz_1 dz_2}{(z_1+z_2)\sin(\pi z_1)\sin(\pi z_2)}\frac{\Gamma(z_1+z_2+1)}{\Gamma(z_1+1)\Gamma(z_2+1)}\exp\left[\lambda z_1 z_2(z_1+z_2)\right] \end{aligned} \tag{13.60}$$

where the integrations over $z_{1,2}$ both go around the contour in the complex plane shown in Fig. 13.1(a). Shifting the contour of integration to the position shown in Fig. 13.1(b), and redefining the integration variables: $z_{1,2} \to \lambda^{-1/3} i x_{1,2}$ in the limit $\beta \to \infty$ ($\lambda^{-1/3} \to 0$) one gets:

$$F_{rsb} = \frac{1}{\beta}\frac{\lambda^{1/3}}{2\pi^2}\int\int_0^{+\infty} dx_1 dx_2 \left[\frac{\sin(x_1x_2(x_1+x_2))}{x_1x_2(x_1+x_2)} + \frac{\sin(x_1x_2(x_1-x_2))}{x_1x_2(x_1-x_2)}\right] \tag{13.61}$$

Substituting the value of $\lambda = \beta^3 g^2/2\mu$ we finally get the result for the zero-temperature free energy:

$$F_{rsb} = \frac{g^{2/3}}{\mu^{1/3}}\frac{\sqrt{3}\Gamma(1/6)}{4\pi^{3/2}} \tag{13.62}$$

To this piece we must now add the replica symmetric contribution. The saddle-point equations have the trivial solution $\phi_a = 0$, with the corresponding energy $E_0 \equiv H_n[\phi_a = 0] = 0$. As we want to get a quantitative result for the constant C_2, we must also include the contribution from the replica fluctuations around this saddle point. This cannot be done just at the level of integrating the quadratic fluctuations. We shall rather make the following (strong) assumption, namely that this whole 'RS' part of the free energy, including the replica fluctuations, is given by the Gaussian replica variational method [83, 93, 94] (note that the Gaussian variational method involves the Gaussian integration over replica fields that in a sense is 'symmetric' with respect to the point $\phi_a = 0$). In the end the hypothesis is best supported by the good result one gets for C_2.

Denoting the Gaussian variational contribution by F_{rv}, the conjecture is that $F = F_{\text{rv}} + F_{\text{rsb}}$. According to Eq. (13.48):

$$\overline{\langle \phi^2 \rangle} = 2\frac{\partial F}{\partial \mu} = \overline{\langle \phi^2 \rangle}_{\text{rv}} - \frac{g^{2/3}}{\mu^{4/3}} \frac{\Gamma(1/6)}{2\sqrt{3}\pi^{3/2}} \tag{13.63}$$

Using the result of [93] for the value of $\overline{\langle \phi^2 \rangle}_{\text{rv}}$ we finally get:

$$\overline{\langle \phi^2 \rangle} = \frac{g^{2/3}}{\mu^{4/3}} \left(\frac{3}{(4\pi)^{1/3}} - \frac{\Gamma(1/6)}{2\sqrt{3}\pi^{3/2}} \right) \simeq 1.00181 \frac{g^{2/3}}{\mu^{4/3}} \tag{13.64}$$

This result can be compared with some numerical simulations of the problem. The scaling in μ and g is obviously correct, the only point to check is the prefactor C_2. Choosing for instance the values of the parameters $\mu = 1$ and $g = 2\sqrt{\pi}$ (when the replica variational method gives $\overline{\langle \phi^2 \rangle}_{\text{rv}} = 3$), from the analytical prediction (13.64) one gets: $\overline{\langle \phi^2 \rangle} \simeq 2.3291$. The result of the numerical simulation with these same values of μ and g is $\overline{\langle \phi^2 \rangle} \simeq 2.45 \pm 0.02$ [88]. Thus, the value predicted by the replica saddle-point summation is rather close, although there is a clear small discrepancy, which is not surprising because we did not take into account the contribution from the replica fluctuations near the RSB saddle point (13.56) (more surprising is that this discrepancy is small!).

Clearly, the vector type of RSB that we have been using on all these zero-dimensional problems is somewhat strange, and at the moment we cannot assert that we control all of its aspects. However, in all these cases, and in particular in the non-trivial case of the toy model, we have obtained good results using this simple recipe. Therefore we now turn to its application to more elaborate problems, starting with systems in one dimension.

13.2 Directed polymers in a random potential

The problem of a directed polymer in a random potential (see Chapter 12) is an important problem that has been much studied recently [16]. Although the situation in a (1+1) dimension, with a delta correlated potential, is relatively well understood, there are still a lot of uncertainties about more complicated cases.

Let us consider again a one-dimensional case with long-range correlations of the potential (see Section 12.1). It is described by a one-dimensional scalar field system with the following Hamiltonian:

$$H = \int_0^L dx \left[\frac{1}{2} \left(\frac{d\phi(x)}{dx} \right)^2 + v(\phi, x) \right] \tag{13.65}$$

where the random potential $v(\phi, x)$ is described by the Gaussian distribution with *non-local* correlations with respect to the fields ϕ:

$$\overline{v(\phi, x)v(\phi', x')} = \delta(x - x') \left[(\text{const}) - g|\phi - \phi'|^{2\alpha} \right] \tag{13.66}$$

where $0 < \alpha < 2$. In this case the behavior of the typical deviation of the polymer 'trajectory' $\phi(x)$ from the origin, $\overline{\langle \phi^2 \rangle} \equiv \phi_0^2 \sim L^{2\zeta}$, is believed to be described by the following simple scaling (see Section 12.1):

$$\phi_0 \sim L^{\frac{3}{2(2-\alpha)}} \, g^{\frac{1}{2(2-\alpha)}} \tag{13.67}$$

In this section we will demonstrate how this result can be obtained in the zero-temperature limit in terms of the proposed replica saddle-point method. The replicated Hamiltonian is:

$$H_n = \int_0^L dx \left[\frac{1}{2} \sum_{a=1}^{n} \left(\frac{d\phi_a(x)}{dx} \right)^2 + \frac{1}{2} \beta g \sum_{a,b=1}^{n} (\phi_a(x) - \phi_b(x))^{2\alpha} \right] \tag{13.68}$$

Strictly speaking, the systematic way of solving this problem according to our general method is the following: one must find n saddle-point trajectories $\phi_a(x)$ for fixed n boundary conditions $\phi_a(L)$, then one has to derive the corresponding energy $\tilde{H}_n[\phi_a(L)]$, and finally one has to find the saddle-point solutions with respect to the values of $\phi_a(L)$.

Here we shall follow a much simpler strategy. Because it is obvious that there always exists the trivial solution $\phi(x) \equiv 0$, we will suppose that the correct scaling can be obtained simply by taking into account one non-trivial saddle-point trajectory. In other words, from the very beginning we are going to look for the saddle-point solutions within the following ansatz:

$$\phi_a(x) = \begin{cases} \phi(x) & \text{for } a = 1, \ldots, k \\ 0 & \text{for } a = k+1, \ldots, n \end{cases} \tag{13.69}$$

Comparing this ansatz with the zero-dimensional exercises of the previous section, we see what it should amount to assuming that the lowest lying configuration dominates. Substituting Eq. (13.69) into the replica Hamiltonian (13.68) in the limit $n \to 0$ one gets:

$$H_k = k \int_0^L dx \left[\frac{1}{2} \left(\frac{d\phi(x)}{dx} \right)^2 - \beta k g \phi^{2\alpha}(x) \right] \tag{13.70}$$

As usual (see the previous section) the free energy is defined by the series:

$$F(L) \sim -\frac{1}{\beta} \sum_{k=1}^{\infty} \frac{(-1)^{k-1}}{k} \exp(-\beta H_k) \tag{13.71}$$

where the value H_k is defined by the corresponding saddle-point solution for $\phi(x)$.

The saddle-point trajectory is defined by the following differential equation:

$$\frac{d^2\phi}{dx^2} = -2\alpha\beta k g \phi^{2\alpha-1} \tag{13.72}$$

with the boundary conditions: $\phi(0) = 0$ and $\phi(L) = \phi_0$. This equation can be easily integrated:

$$\int_0^{\phi(x)} \frac{d\phi}{\sqrt{\lambda - \phi^{2\alpha}}} = x\sqrt{2\beta k g} \tag{13.73}$$

where the integration constant λ is defined by the boundary condition:

$$\int_0^{\phi_0} \frac{d\phi}{\sqrt{\lambda - \phi^{2\alpha}}} = L\sqrt{2\beta k g} \tag{13.74}$$

Substituting this solution into the Hamiltonian (13.70), we obtain after some simple algebra:

$$H_k = k \left[-\beta k g \lambda L + \sqrt{2\beta k g} \int_0^{\phi_0} d\phi \sqrt{\lambda - \phi} \right] \tag{13.75}$$

Taking the derivative of H_k with respect to ϕ_0 (and taking into account the constraint (13.74)) one finds the following saddle-point solution:

$$\phi_0 \sim L^{\frac{1}{1-\alpha}} (\beta k g)^{\frac{1}{2(1-\alpha)}} \tag{13.76}$$

and $\lambda = \phi_0^{2\alpha}$. Its energy (13.75) is:

$$H_k = -\frac{(\text{const})}{\beta} (\beta k)^{\frac{2-\alpha}{1-\alpha}} L^{\frac{1+\alpha}{1-\alpha}} g^{\frac{1}{1-\alpha}} \tag{13.77}$$

Now we proceed as before, introducing an integral representation of the series (13.71):

$$F(L) = \frac{1}{2i\beta}\int_C \frac{dz}{z\sin(\pi z)} \exp\left[(\text{const})(\beta z)^{\frac{2-\alpha}{1-\alpha}} L^{\frac{1+\alpha}{1-\alpha}} g^{\frac{1}{1-\alpha}}\right] \tag{13.78}$$

where the integration goes over the contour in the complex plane shown in Fig. 13.1(a). Rescaling of the integration variable:

$$z \to \frac{ix}{\beta} L^{-\frac{1+\alpha}{2-\alpha}} g^{-\frac{1}{2-\alpha}} \tag{13.79}$$

in the limit $\beta \to \infty$ we get the following scaling of the free energy (13.78):

$$F(L) \sim L^{\frac{1+\alpha}{2-\alpha}} g^{\frac{1}{2-\alpha}} \tag{13.80}$$

The scaling of ϕ_0 as a function of L can be obtained either from the general relation $\phi_0 \sim F(L)L$, or directly from the saddle-point value (13.76) after rescaling of the factor (βk) (which in the integral representation turns into (βz)) according to Eq. (13.79):

$$\phi_0(L) \sim L^{\frac{3}{2(2-\alpha)}} g^{\frac{1}{2(2-\alpha)}} \tag{13.81}$$

which coincides with the result (13.67) given by the naive energy arguments, as well as by more elaborate calculations.

Although the example demonstrated in this section provides no new results one can hope that the proposed method could turn out to be also useful when applied for other types of directed polymer model, and in particular in larger dimensions.

13.3 Replica instantons and Griffiths singularities in the disordered Ising Model

According to the original statement of Griffiths [44], the free energy of the random Ising ferromagnet in the temperature interval above its ferromagnetic phase transition point T_c and below the critical point $T_c^{(0)}$ of the corresponding pure system must be a non-analytic function of the external magnetic field h, such that in the limit $h \to 0$ the free energy as a function of h has essential singularity. Because this type of phenomenon, namely, existence of non-analytic non-perturbative contributions to thermodynamical functions in random systems, seems to be a rather general one, at present it has become common to call any such contribution the 'Griffiths singularity'.

Owing to intensive theoretical [45] and numerical [46] studies of the Griffiths singularities it was also discovered that the *dynamical* properties of the system in the temperature interval $T_c < T < T_c^{(0)}$ are not just ordinary paramagnetic properties. According to numerical simulations the time correlation functions here can be described by the so-called stretched-exponential asymptotic behavior ($\sim \exp\{-(\text{const})t^{\lambda(T)}\}$ with $\lambda(T) < 1$) which is much slower than the usual exponential one, as it should be in the paramagnetic phase. On the other hand, recent analytical calculations [95] yield a time decay of the form $\exp\{-(\text{const})(\ln t)^{D/(D-1)}\}$. In any case, to underline that the properties of the system in the temperature interval $T_c < T < T_c^{(0)}$ are not quite paramagnetic, it has become common to call the state of the system here the 'Griffiths phase'.

At the level of 'hand-waving arguments', the dynamical Griffiths phenomena can be explained 'theoretically' rather easily: considering, for example, the bond diluted Ising model, one can note that at temperatures below $T_c^{(0)}$ in the 'ocean' of the zero magnetization paramagnetic background the random system must contain disconnected locally ordered 'ferromagnetic islands' (composed only of the pure system bonds) of all sizes, which, in turn, creates the whole spectrum (up to infinity) of relaxation times. Having an infinite spectrum of relaxation times, with some imagination, it is not difficult to derive any relaxation law one likes, and the stretched-exponential one in particular.

Although it is commonly believed that the main point of the above 'explanation', namely the existence of an infinite number of local minima states, must be a general key for understanding the Griffiths phenomena (both dynamical and statistical mechanical), despite many efforts during the past 30 years, it has turned out to be extremely difficult to construct a more elaborate or convincing theory. For that reason any progress in understanding of the effects produced by numerous local minima states (which, so to say, are away from the perturbative region) looks valuable.

In this section I am going to demonstrate the origin of non-perturbative contributions to the thermodynamical functions of a weakly disordered (random temperature) D-dimensional ($D < 4$) Ising ferromagnet in the zero magnetic field in the paramagnetic phase away from the critical point [96]. In the continuous limit this system can be described by the usual Ginsburg–Landau Hamiltonian:

$$H[\phi(x);\delta\tau(x)] = \int d^D x \left[\frac{1}{2}(\nabla\phi(x))^2 + \frac{1}{2}(\tau - \delta\tau(x))\,\phi^2(x) + \frac{1}{4}g\phi^4(x)\right] \quad (13.82)$$

Here, as usual, the quenched disorder is described by random spatial fluctuations of the local transition temperature $\delta\tau(x)$ whose probability distribution is taken to be symmetric and Gaussian:

$$P[\delta\tau] = p_0 \exp\left(-\frac{1}{4u}\int d^D x(\delta\tau(x))^2\right), \tag{13.83}$$

where $u \ll g$ is the small parameter that describes the strength of the disorder, and p_0 is an irrelevant normalization constant. For notational simplicity, I define the sign of $\delta\tau(x)$ in Eq. (13.82) so that positive fluctuations lead to locally ordered regions, whose effects will be the object of our further study.

To avoid possible misunderstandings, I would like to note that the problem considered here is actually rather far from the original one studied by Griffiths as well as by many other people since. Because the shift of T_c in the weakly disordered ferromagnet compared to $T_c^{(0)}$ of the pure system is of the order of $\sqrt{u}$, in the limit $u \ll g$ the interval of temperatures $T_c < T < T_c^{(0)}$ where the so-called Griffiths phase is expected to take place, appears to be well inside of the temperature interval $\tau_g \sim g^{2/(4-D)}$ where the critical fluctuations are essential, and where the mean-field approach considered here can not be used. For that reason, in the considered range of temperatures $\tau \gg \tau_g$ it is hardly reasonable to look for non-analytic behavior of the free energy as a function of the external magnetic field (at least the present approach in terms of the replica instantons modified by the external field h does not seem to indicate any non-analyticity in h). The aim of the present consideration is just to demonstrate that in addition to the 'usual' Griffiths singularities in terms of the external field, the free energy of the random ferromagnet (in the zero magnetic field) must also be non-analytic in the value of the parameter which describes the strength of the disorder.

As far as the corresponding pure system ($u \equiv 0$) is concerned (see Chapter 7), it is well known that in the close vicinity of T_c, at $|\tau| \ll \tau_g \sim g^{2/(4-D)}$, its properties are defined by non-Gaussian critical fluctuations (which can be studied, for example, in terms of the ϵ-expansion renormalization group approach), while away from T_c, at $|\tau| \gg \tau_g$, the situation becomes Gaussian, and everything becomes very simple. Here the total magnetization is defined by the order parameter $\langle\phi\rangle \equiv \phi_0(\tau)$, which is equal to 0 above T_c, and is equal to $\pm\sqrt{|\tau|/g}$ below T_c; the asymptotic behavior of the correlation function $G(x-x') \equiv (\langle\phi(x)\phi(x')\rangle - \phi_0^2)$ is defined only by the Gaussian fluctuations: $G(x) \sim |x|^{-(D-2)}$; and the singular part of the free energy $f(\tau)$ scales with the temperature as $f(\tau) \simeq \tau^{D/2}$.

Usually, the random system, defined by the Hamiltonian (13.82), was studied from the point of view of the effects produced by the quenched disorder on the critical phenomena in the close vicinity of the phase transition point. A renormalization-group consideration shows that if the temperature is not too close to T_c, at $\tau_u \ll \tau \ll \tau_g$ (where the disorder-dependent crossover temperature scale $\tau_u \sim u^{1/\alpha}$ is defined by the specific heat critical exponent $\alpha > 0$ of the pure system) the critical behavior is essentially controlled by the pure system fixed point, and the disorder produces only irrelevant corrections. On the other hand, in the close vicinity of the critical point, at $\tau \ll \tau_u$, the critical behavior moves into a new universality class defined by the so-called random fixed point, which turns out to be universal (Chapter 8). In recent years, however, this very nice physical picture has been questioned on the grounds that the renormalization-group approach completely misses the presence of numerous local minima configurations of the random Hamiltonian (13.82), which, in principle, may cause the spontaneous replica symmetry breaking in the interaction parameters of the critical fluctuations, which, in turn, may ruin the above physical scenario (Chapter 9).

Leaving aside the discussion of this very difficult problem, here I would like to pose a much more simple question: how do the thermodynamic functions of this system depend on the strength of the disorder u (in the limit $u \to 0$) far way from T_c, at $\tau \gg \tau_g$, where the behavior of the pure system is correctly described by the Gaussian approximation? It turns out that even this, which seems an almost trivial question, is not so easy to answer.

Of course, first of all, one can proceed in a straightforward way, developing the perturbation theory in powers of the parameter u at the background of the pure system paramagnetic state $\langle\phi\rangle = 0$ by using the Gaussian approximation for the thermal fluctuations. There is nothing wrong with this approach, but the problem is that it can not give *all* thermodynamic contributions which exist at $u \neq 0$. The drawback of this type of perturbation theory is the same as that of the renormalization group: it completely misses the existence of numerous (macroscopic number) local minima configurations of the random Hamiltonian (13.82).

At the level of 'hand-waving arguments' it is very easy to see what all these off-perturbative states are. At any $u \neq 0$ there exists a finite (exponentially small) density of 'ferromagnetic islands' in which the local (random) temperature is below T_c (such that $\delta\tau(x) > \tau$), and the minimum energy configurations here are achieved at a non-zero local value of the order parameter: $\phi_0(x) \sim \pm\sqrt{(\delta\tau - \tau)/g}$. Because the spatial density of such islands is finite, and each island provides two ($\pm$) possibilities for the local magnetization, the total number of the local minima configurations in the system must be exponential in its volume.

Formally, to take into account the contributions of all these states, one has to proceed as follows. For a given arbitrary quenched function $\delta\tau(x)$ one has to find all possible local minima solutions of the saddle-point equation:

$$-\Delta\phi(x) + (\tau - \delta\tau(x))\phi(x) + \phi(x)^3 = 0 \tag{13.84}$$

Then one has to substitute these solutions into the Hamiltonian (13.82) and calculate the corresponding thermodynamic weights. Next, to compute the partition function one has to perform a summation over all the solutions, and finally to get the corresponding free energy one has to take the logarithm of the partition function and average it over random functions $\delta\tau(x)$ with the probability distribution (13.83). Clearly, it is hardly possible that such a program can be implemented.

On the other hand, as usual, for the systems that contain quenched disorder we can use the standard replica method and reduce the problem of the quenched averaging to the annealed one for n copies of the original system:

$$F = -\lim_{n\to 0} \frac{1}{n} \left(\overline{Z^n} - 1\right) \tag{13.85}$$

where:

$$\overline{Z^n} = \prod_{a=1}^{n} \left[\int \mathcal{D}\phi_a(x)\right] \exp\left(-H_n[\phi_a(x)]\right) \tag{13.86}$$

is the replica partition function and

$$H_n[\phi_a(x)] = \int d^D x \left[\frac{1}{2}\sum_{a=1}^{n}(\nabla\phi_a)^2 + \frac{1}{2}\tau\sum_{a=1}^{n}\phi_a^2 + \frac{1}{4}g\sum_{a=1}^{n}\phi_a^4 - \frac{1}{4}u\sum_{a,b=1}^{n}\phi_a^2\phi_b^2\right] \tag{13.87}$$

is the *spatially homogeneous* replica Hamiltonian.

Now, if we are intending to take into account non-trivial local minima states, instead of solving the original inhomogeneous stationary equation (13.84), we can consider the corresponding replica saddle-point equations:

$$-\Delta\phi_a(x) + \tau\phi_a(x) + \phi_a^3(x) - u\phi_a(x)\sum_{b=1}^{n}\phi_b^2(x) = 0 \tag{13.88}$$

Because until now all the transformations were exact, these equations must contain (maybe in a slightly hidden way) all the relevant non-trivial states, which in the language of the original random Hamiltonian correspond to rare ferromagnetic islands.

At this stage we can note one very simple point. Looking for various types of solutions of the above equations one can try first of all the simplest possible 'replica symmetric' ansatz, in which the fields in all replicas are assumed to be equal: $\phi_a(x) = \phi(x)$. In this case the last term in the Eqs. (13.88) (which contains the factor $\sum_{b=1}^{n} \phi_b^2(x) = n\phi^2(x)$) drops away in the limit $n \to 0$, and these equations reduce to the *pure system* saddle-point equation:

$$-\Delta\phi(x) + \tau\phi(x) + \phi(x)^3 = 0 \tag{13.89}$$

which at $\tau > 0$ has only trivial solution $\phi(x) \equiv 0$. This means that in any non-trivial solution of the Eqs. (13.88) the fields $\phi_a(x)$ in different replicas *cannot* be all equal. In other words, the symmetry among replicas in the *replica vector* $\phi_a(x)$ must be broken.

Following the general strategy developed in this chapter, let us assume that in addition to the trivial replica symmetric (RS) solutions of the saddle-point equations (13.88), there exist other types of solution, which are *well separated* in the configurational space from the RS state. In this case, denoting the contribution of these non-trivial states by the label 'replica symmetry breaking' (RSB), the replica partition function, Eq. (13.86), can be decomposed into two parts:

$$\overline{Z^n} = Z_{\text{rs}} + Z_{\text{rsb}} \tag{13.90}$$

where Z_{rs} contains all the perturbative contributions in the vicinity of the trivial state $\phi_a(x) = 0$. As usual, this partition function can eventually be represented in the form:

$$Z_{\text{rs}} = \exp(-nVf_{\text{rs}}) \tag{13.91}$$

where V is the volume of the system, and f_{rs} is the free energy density, which contains the pure system leading term $\sim \tau^{D/2}$ (at temperatures not too close to T_c, $\tau \gg \tau_g$), plus the perturbation series in powers of the disorder parameter u.

Thus, in terms of the general replica approach, according to Eq. (13.85) for the total free energy we get:

$$F = Vf_{\text{rs}} + F_{\text{rsb}} \tag{13.92}$$

where the additional RSB part of the free energy

$$F_{\text{rsb}} = -\lim_{n\to 0} \frac{1}{n} Z_{\text{rsb}} \tag{13.93}$$

must contain all non-perturbative contributions (if any) that are away from the trivial state $\phi_a = 0$. It is this part of the free energy that will be the point of our further study.

The simplest possible non-trivial replica structure for the solutions of the saddle-point equations (13.88) can be taken in the following form:

$$\phi_a(x) = \begin{cases} \pm\phi(x) & \text{for } a = 1, \ldots, k \\ 0 & \text{for } a = k+1, \ldots, n \end{cases} \tag{13.94}$$

where k is the integer value parameter $k = 1, 2, \ldots, n$, which defines a given structure of the trial replica vector ϕ_a (note that the value $k = 0$ should be excluded because it describes the trivial RS solution, which is already taken into account in f_{rs}). The solutions in Eq. (13.94) are taken with '$\pm$' signs, because the saddle-point equations (13.88) are invariant with respect to the global change of signs of the replica fields. Substituting this anzatz into Eqs. (13.88) as well as into the replica Hamiltonian (13.87), one finds that for a given value of the parameter k the fields $\phi(x)$ in Eq. (13.94) are defined by the solutions of the following saddle-point equation:

$$-\Delta\phi(x) + \tau\phi(x) - \lambda(k)\phi(x)^3 = 0 \tag{13.95}$$

and the thermodynamic weight of any such solution is defined by the energy:

$$E(k) = k \int d^D x \left[\frac{1}{2}(\nabla\phi(x))^2 + \frac{1}{2}\tau\phi^2(x) - \frac{1}{4}\lambda(k)\phi^4(x)\right] \tag{13.96}$$

where

$$\lambda(k) = (uk - g) \tag{13.97}$$

As usual, the free energy is given by the series:

$$F_{\text{rsb}} \sim -\sum_{k=1}^{\infty} \frac{(-1)^{k-1}}{k} \exp(-E(k)) \tag{13.98}$$

where the value of $E(k)$ is defined by the corresponding saddle-point solution of Eq. (13.95).

At this stage we see that the situation becomes rather different from those studied in the previous sections. If we choose the obvious space-independent solution $\phi = (\tau/\lambda(k))^{1/2}$, we would find that the value of $E(k)$ is proportional to the volume V of the system: $E(k) = -\frac{1}{4}k(\tau^2/\lambda(k))\,V$. Then, the summation of the series (13.98) would immediately yield the contribution of the free energy of the order of $\exp\{-(\tau^2/4u)V\}$, which is not proportional to the volume of the system. Therefore this type of solution (as well as any other solution with an energy $E(k)$ proportional to the volume) is irrelevant for the bulk properties.

Thus, we have to look for *localized* solutions: the ones which are local in space (breaking translation invariance) and which have *finite* energy. Let us suppose that such an 'instanton'-type solution exists (see below), and that for a given k it is characterized by the spatial size $R(k)$. Then, if we take into account only the one-instanton contribution (or in other words we consider a gas of *non-interacting* instantons), owing to the trivial entropy factor V/R^D (this is the number of positions of the object of size R in the volume V) we get a free energy proportional to the volume:

$$F_{\text{rsb}} \sim -\sum_{k=1}^{\infty} \frac{(-1)^{k-1}}{k} \frac{V}{R(k)^D} \exp\{-E(k)\} \tag{13.99}$$

where $E(k)$ must be finite (volume independent).

Coming back to the saddle-point equation (13.95), let us consider the range of the parameter k such that $\lambda(k) = (uk - g) > 0$ (i.e. $k > [g/u]$). Rescaling the fields:

$$\phi(x) = \sqrt{\frac{\tau}{\lambda(k)}}\psi(x\sqrt{\tau}) \tag{13.100}$$

instead of Eq. (13.95) one obtains the following differential equation which contains no parameters:

$$-\Delta\psi(z) + \psi(z) - \psi^3(z) = 0 \tag{13.101}$$

Correspondingly, for the energy, (Eq. (13.96)), one obtains:

$$E(k) = \frac{k}{uk - g}\tau^{(4-D)/2} E_0 \tag{13.102}$$

where

$$E_0 = \int d^D z \left[\frac{1}{2}(\nabla\psi(z))^2 + \frac{1}{2}\psi^2(z) - \frac{1}{4}\psi^4(z)\right] \tag{13.103}$$

Equation (13.101) is well know in field theory (see e.g. [97]). It can be shown that in dimensions $D < 4$ this equation has spherically symmetric instanton-like solutions such that:

$$\begin{aligned} &\psi(|z| \le 1) \simeq \psi(0) \sim 1, \\ &\psi(|z| \gg 1) \sim \exp(-|z|) \to 0. \end{aligned} \tag{13.104}$$

The energy, Eq. (13.103), of such a solution is a finite and *positive* number. Of course, for a *generic* value of the field $\psi(0)$ at the origin, the solution tends to the values $\psi(|z| \to \infty) = \pm 1$ which are the extrema of the potential $\frac{1}{2}\psi^2 - \frac{1}{4}\psi^4$, and any such solution has divergent energy (13.103). However, there exists a discrete set of initial values ψ_0 for which the solutions (exponentially) tend to zero at infinity, and which have finite energies. It can be shown that the solution with the minimal energy E_0 corresponds to the minimal value of $|\psi_0|$ in the set. In particular, at $D = 3$, $\psi_0 \simeq 4.34$ and $E_0 \simeq 18.90$. For our further calculations with the exponential accuracy it will be sufficient to take into account only the solution with the minimal energy.

According to the rescaling (13.100), in terms of the original fields $\phi(x)$ the size of the instanton is $R = \tau^{-1/2}$ (note that it does not depends on k), which coincides with the usual correlation length of the Ginsburg–Landau theory. Substituting this value of R as well as the energy (13.102) of the instanton into the series (13.99) for the free energy one gets:

$$F_{\text{rsb}} \simeq -V\tau^{D/2} \sum_{k>[g/u]}^{\infty} \frac{(-1)^{k-1}}{k} 2^k \exp\left[-E_0 \frac{k}{uk-g} \tau^{(4-D)/2}\right] \tag{13.105}$$

(the factor 2^k appears because of independent summation over $\pm$ signs, in k non-zero replicas, Eq. (13.94)). It can be easily shown that under the considered conditions on the parameters u, g and τ ($u \ll g \ll 1$, and $g^{2/(4-D)} \ll \tau \ll 1$) the leading contribution in the above series with exponential accuracy comes from the region $k \gg g/u \gg 1$:

$$\frac{1}{V} F_{\text{rsb}} \simeq \tau^{D/2} \exp\left[-E_0 \frac{\tau^{(4-D)/2}}{u}\right] \times \sum_{k \gg g/u}^{\infty} \frac{(-1)^{k-1}}{k} 2^k \tag{13.106}$$

Here the absolute value of the series $\sum_{k=k_o \gg 1}^{\infty} k^{-1}(-1)^{k-1} 2^k$ can be estimated by the upper bound $\sim k_o^{-1} 2^{k_o}$, and since it is assumed that $\tau \gg g^{2/(4-D)}$ the term $(g/u)\ln 2$, which appears in the exponential, can be dropped in comparison with $E_0 \tau^{(4-D)/2}/u$. Thus, for the density of the free energy we finally obtain the following contribution:

$$\frac{1}{V} F_{\text{rsb}} \sim \exp\left[-E_0 \frac{\tau^{(4-D)/2}}{u}\right] \tag{13.107}$$

(where we drop all pre-exponential factors, which within the present accuracy of calculations can not be defined).

Now we have to investigate under which conditions the effects of the critical fluctuations are irrelevant for the mean-field approximation considered above. Note first of all, that one should not be confused by the 'wrong' sign of the ϕ^4-interaction term in the energy function (13.96), which for the usual field theory would indicate its absolute instability. Here, as usual in the replica theory, in the limit $n \to 0$ everything turns 'upside down', so that the minima of the physical free energy actually correspond to the *maxima* of the replica free energy. It can easily be shown (see below) that formal integration over n-component replica fluctuations around a considered instanton solution in the limit $n \to 0$ yields a physically sensible result.

Proceeding the same way as in the usual Ginsburg–Landau theory (Section

7.1), let us introduce small fluctuations $\varphi_a(x)$ near the instanton solution, Eqs. (13.94) and (13.100): $\phi_a(x) = \phi_a^{(\text{inst})}(x) + \varphi_a(x)$. In the Gaussian approximation we get the following Hamiltonian for the fluctuating fields:

$$H[\varphi] = \int d^D x \left[\frac{1}{2} \sum_{a=1}^{n} (\nabla \varphi_a(x))^2 + \frac{1}{2} \tau \sum_{a,b=1}^{n} T_{ab}(x) \varphi_a(x) \varphi_b(x) \right] \tag{13.108}$$

where the matrix $T_{ab}(x)$ contains the $k \times k$ block:

$$T_{ab}^{(k)}(x) = \left(1 - \frac{uk - 3g}{uk - g} \psi^2(x\sqrt{\tau}) \right) \delta_{ab} - \frac{2u}{uk - g} \psi^2(x\sqrt{\tau}) \tag{13.109}$$

(where $a, b = 1, \dots, k$) and the diagonal elements for the remaining $(n - k)$ replicas:

$$T_{ab}^{(n-k)} = \left(1 - \frac{uk}{uk - g} \psi^2(x\sqrt{\tau}) \right) \delta_{ab} \tag{13.110}$$

(where $a, b = k + 1, \dots, n$). Here the function $\psi(z)$ is the instanton solution, Eq. (13.104).

Because the mass term in the Hamiltonian (13.108) is proportional to τ, the behavior of the correlation function of the fluctuating fields at scales $|x| \ll R_c \sim \tau^{-1/2}$ appears to be the same as in the Ginsburg–Landau theory: $G_{ab}(x - x') = \langle \varphi_a(x) \varphi_b(x') \rangle \sim |x - x'|^{-(D-2)} \delta_{ab}$ (beyond R_c this correlation function decays exponentially). Therefore, the typical value of the fluctuations $\langle \varphi^2 \rangle$ can be estimated in the usual way:

$$\langle \varphi^2 \rangle \sim \frac{1}{n} \sum_{a=1}^{n} R_c^{-D} \int_{|x|<R_c} d^D x G_{aa}(x) \sim \tau^{(D-2)/2} \tag{13.111}$$

The saddle-point approximation considered above is justified only if the typical value of the fluctuations is small compared to the value of the 'background' instanton field $\phi^{(\text{inst})}(x) \sim \sqrt{\tau / \lambda(k)}$ (see Eq. (13.100)):

$$\tau^{(D-2)/2} \ll \frac{\tau}{\lambda(k)} \Rightarrow \lambda(k) \sim uk \ll \tau^{(4-D)/2} \tag{13.112}$$

On the other hand, the contribution (13.107) appears from summation in the region $k \gg g/u$. Thus, one can get this type of contribution to the free energy only in the following interval of summation over k:

$$\frac{g}{u} \ll k \ll \frac{1}{u} \tau^{(4-D)/2} \tag{13.113}$$

This interval exists provided

$$\tau \gg g^{2/(4-D)} \tag{13.114}$$

which is the usual Ginsburg–Landau criterion.

One can also arrive at the same conclusion deriving the fluctuational contribution to the RSB part of the free energy by direct integration over fluctuating fields using the Gaussian Hamiltonian (13.108) (this way one can also check that this contribution contains no imaginary parts which would happen, if the considered extrema correspond to physically unstable field configuration). Assuming the θ-like structure of the instanton solution: $\psi(|z| \leq 1) \simeq \psi(0) \equiv \psi_o \sim 1$ and $\psi(|z| > 1) = 0$, the fluctuating modes with momenta $p \ll \sqrt{\tau}$ and $p \gg \sqrt{\tau}$ in the Hamiltonian (13.108) can be explicitly decoupled:

$$H = \frac{1}{2}\sum_{a,b=1}^{n} \int_{|p| \gg \sqrt{\tau}} \frac{d^D p}{(2\pi)^D} \left[p^2 \delta_{ab} + \tau T_{ab}\right] \varphi_a(p)\varphi_b(-p) + \frac{1}{2}\sum_{a=1}^{n} \int_{|p| \ll \sqrt{\tau}} \frac{d^D p}{(2\pi)^D} p^2 |\varphi_a(p)|^2 \quad (13.115)$$

where the p-independent matrix T_{ab} is given by Eqs. (13.109)–(13.110), in which instead of the function $\psi(x\sqrt{\tau})$ one has to substitute the constant ψ_o.

The integration over the *replica symmetric* modes with momenta $p \ll \sqrt{\tau}$ (they correspond to fluctuations at scales much bigger than the size of the instanton) described by the second term of the Hamiltonian (13.115), gives the contribution of the form $\exp(-nV\tilde{f}_{\text{rs}})$, and it vanishes in the limit $n \to 0$ (note that in the RSB part of the free energy we have to keep only the terms that remain *finite* in the limit $n \to 0$ and not linear in n). This is natural, because this contribution is already contained in the RS part of the free energy.

The integration over the modes with momenta $p \gg \sqrt{\tau}$ is slightly cumbersome but straightforward:

$$\tilde{Z}_{\text{rsb}} \equiv \prod_{p \gg \sqrt{\tau}} \left[\int \mathcal{D}\varphi_a(p)\right] \exp\{-H[\varphi_a(p)]\}$$

$$= \exp\left[-\frac{1}{2}\tau^{-D/2} \int_{p \gg \sqrt{\tau}} d^D p \operatorname{Tr} \ln\left(p^2 \delta_{ab} + \tau T_{ab}\right)\right] \quad (13.116)$$

The matrix under the logarithm in the above equation contains $(k-1)$ eigenvalues:

$$\lambda_1 = p^2 + \tau\left(1 - \frac{uk - 3g}{uk - g}\psi_o^2\right) \quad (13.117)$$

one eigenvalue:

$$\lambda_2 = p^2 + \tau\left(1 - \frac{uk - 3g}{uk - g}\psi_o^2\right) - \tau\frac{2uk}{uk - g}\psi_o^2 \quad (13.118)$$

and $(n-k)$ eigenvalues:

$$\lambda_3 = p^2 + \tau\left(1 - \frac{uk}{uk-g}\psi_o^2\right) \tag{13.119}$$

Substituting these eigenvalues into Eq. (13.116), after simple algebra in the limit $n \to 0$ one eventually obtains the following result:

$$\tilde{Z}_{rsb} \sim \exp\left(\frac{3k}{2(uk-g)} g\psi_o^2\right) \tag{13.120}$$

Thus we see that in the region $\tau \gg g^{2/(4-D)}$ the factor $kg/(uk-g)$ in the exponential of the above equation is small compared to the leading term $k\tau^{(4-D)/2}/(uk-g)$ given by the saddle-point solution, Eq. (13.102).

It is interesting to note that a non-analytic instanton contribution of the form given by Eq. (13.107) can be easily 'derived' based on qualitative physical arguments. Let us again consider the random Hamiltonian (13.82) at temperatures above T_c ($\tau > 0$), and let us estimate the contribution to the free energy coming from rare 'ferromagnetic islands' where $\delta\tau(x) > \tau$. In the mean-field regime at finite values of τ the typical smallest (most probable) size of such islands is $R_c \sim \tau^{-1/2}$. Therefore, according to the probability distribution, (Eq. (13.83), in the limit of weak disorder ($u \to 0$) the contribution of the islands to the free energy with the exponential accuracy can be estimated by their probability:

$$\begin{aligned}\delta F &\sim \int_\tau^\infty d(\delta\tau) \exp\left(-\frac{(\text{const})}{u} R_c^D (\delta\tau)^2\right) \\ &\sim \int_\tau^\infty d(\delta\tau) \exp\left(-\frac{(\text{const})}{u} \tau^{-D/2} (\delta\tau)^2\right) \\ &\sim \exp\left(-\frac{(\text{const})}{u} \tau^{(4-D)/2}\right)\end{aligned} \tag{13.121}$$

which (up to the undefined (const) factor) coincides with the result (13.107).

Of course, exponentially small contributions to the free energy (as well as to other thermodynamical functions) of the type (13.107) are not so important for thermodynamical properties of the random ferromagnet in the considered paramagnetic temperature region. Nevertheless, the fact of their existence seems very interesting for two reasons.

First, it tells us that even in the mean-field regime the free energy of the random ferromagnet must be a non-analytic function of the parameter that describes the strength of disorder $u \to 0$, which is interesting in itself.

Second, it indicates the importance of non-linear excitations, which in terms of the present replica field theoretical approach are described by the localized instanton-like solutions of the stationary equations. In the considered mean-field region away from T_c these excitations provide only exponentially small corrections. However, in the close vicinity of the critical point the presence of instantons (which is ignored in the standard renormalization-group approach), and their interactions with the critical fluctuations may produce dramatic effects on the critical properties of the phase transition. It is worth noting that although in the scaling regime (at $T = T_c$) the situation looks very different from that considered here, the corresponding stationary equations (13.88) (with $\tau = 0$) also have non-linear instanton-like solutions with the RSB structure given by Eq. (13.94). One can easily check that in the dimension $D = 4$ these solutions can be found explicitly [98]:

$$\phi(x) = \sqrt{\frac{8}{(uk-g)}}\frac{R}{R^2 + |x|^2} \tag{13.122}$$

where the size of the instanton R appears to be the *zero mode* (the energy of the instanton does not depend on R). In dimensions below but close to four (at $\epsilon = (4 - D) \ll 1$) the field configuration given by Eq. (13.122) can be considered as the approximate solution that contains the parameter R as the *soft mode*, since the energy of the instanton, Eq. (13.96), depends on R very weakly:

$$E(k) = \frac{4}{3} S_D R^{-\epsilon} \frac{k}{uk-g} \tag{13.123}$$

(here S_D is the square of the unite D-dimensional sphere).

At present, however, it is not quite clear how all these non-linear instanton excitations could be incorporated into the self-consistent theory of the critical fluctuations. At the qualitative level one can only note that because the degrees of freedom of this type explicitly break the replica symmetry, this gives some support for the 'heuristic' renormalization-group approach described in Chapter 9, where it was assumed that owing to interactions of the fluctuations with this type of non-perturbative excitation the replica symmetry in the effective matrix, describing non-linear interactions of the fluctuating fields, is spontaneously broken. Hopefully, the study described here can stimulate further much deeper investigation of the physics of the phase transition in random ferromagnets.

In this chapter we have considered a method to analyse random systems by summing up various saddle-point contributions in the replicated Hamiltonian. Hopefully, it may open a new route in this type of study. In particular, the application to finite-dimensional systems, such as directed polymers on one hand, and the random temperature Ising models on the other hand, looks quite interesting. Indeed we have seen in this last case how this method allows us to take into account instanton contributions that are usually out of reach of most analytic methods in these systems. Such instanton contributions have been argued to be important for a long time [75, 78]. The above considerations show that in principle we can get them under control with the present approach.

Clearly this method is still not totally understood. We have pointed out that it involves a single basic rule, stating the way one has to order the various saddle points in replica space. Within this hypothesis it gives reasonable results in all the cases considered, but of course more studies are needed to justify this hypothesis.

14
Conclusions

In this book we have studied various effects produced by quenched disorder on thermodynamical properties of statistical systems. Considering different types of model, the emphasis has been made on the demonstration of the basic theoretical approaches and ideas. Although the considered systems and the corresponding problems involved (such as spin glasses, critical phenomena, directed polymers etc.) may at first look quite different, the aim of the book was to demonstrate that basically all these problems are deeply interconnected. I was trying to convince the reader that to work successfully on any particular problem in this field one needs to be familar with all the methods and ideas of statistical field theory. The physics of both the spin-glass state and critical phenomena in weakly disordered systems involve the ideas of the scaling theory of phase transitions, and the basic concepts of the replica theory of spin glasses. The aim of this book was to take the reader, starting from fundamentals and demonstrating well-established solutions of various problems, to the frontier of modern research. Here we are facing quite a few fundamental problems, both long-standing and new ones, still waiting for their solutions.

The most appealing problem in the scope of spin glasses had remained a question for almost two decades: whether or not the mean-field RSB physical picture (described in Chapters 2–6) is valid, at least at the qualitative level, for realistic spin glasses with finite-range interactions. In the opinion of one part of the physical community (including myself), recent developments in this field, both numerical and analytical, seem to favor the mean-field scenario and not the droplet model [7], although there is another part of physicists which believes just the opposite. At the moment rather hot debates about this issue are well under way [5, 6]. We did not discuss them in this book, because instead of a systematic and slow pedagogical consideration it would make the content too chaotic and inconclusive. Nevertheless, I believe that after studying this book a reader with a fresh mind will become well equipped to investigate this delayed problem and to find its ultimate solution.

Another very important problem in this field (which, in fact, is interconnected with the previous one) concerns the dynamical properties of spin glasses. In this actively developing domain of research, both analytical [8] and experimental [9] (briefly described in Chapter 6), one finds really fascinating phenomena like history dependence and breaking of the fluctuation–dissipation theorem, which go well beyond the framework of the usual non-equilibrium thermodynamics. Of course, the construction of the systematic and selfconsistent theory of such non-equilibrium phenomena turns out to be an extremely difficult problem, and despite great efforts over many years it is still far from being completed.

In the scope of the critical phenomena, according to the results described in Chapter 9, we conclude that spontaneous replica symmetry breaking in the interaction parameters of the critical fluctuations has a dramatic effect on the renormalization-group flows and on the critical properties. Here one finds that for a generic RSB structure there exist no stable fixed points, and the RG flows lead to the *strong coupling regime* at the finite (exponentially large) spatial scale. Physically this could mean that in general (in deep contrast with the usual scaling theory of the phase transitions in homogeneous systems), the critical phenomena in disordered systems should be described in terms of an infinite hierarchy of correlation lengths, which, in turn, would mean breaking of scaling and universality. The crucial problem, however, is that one still has to find convincing arguments demonstrating the origin of the spontaneous RSB in the interactions of the critical fluctuations. Presumably, to create firm ground for this type of theory one has to find a systematic way of taking into account the contributions of the off-perturbative instanton-like solutions of the corresponding mean-field equations. Hopefully this can be done in the framework of the vector replica symmetry breaking scheme described in Chapter 13.

Another general problem that seems to be interconnected with the RSB phenomena discussed above is related to the existence of the Griffiths phase above the ferromagnetic phase transition point. Usually the Griffiths phenomena are associated with the existence of numerous off-perturbative local minima metastable states separated from the true ground state by big energy barriers. On the other hand, in the mean-field theory of spin glasses the RSB phenomena can be interpreted as a factorization of the phase space into a hierarchy of 'valleys' of local minima states, separated by macroscopic energy barriers. If such a general physical interpretation on a qualitative level can also be adopted for weakly disordered systems in the vicinity of the critical point (with the reservation that the values of the corresponding energy barriers are finite), then this qualitative physical picture would fit nicely with the Griffiths phenomena. In connection with this, in the last chapter we have considered a new method to analyze random systems by summing up various saddle-point contributions in the replicated Hamiltonian. Hopefully, it may open a new route in this type of study. In particular, the application to the weakly disordered Ising model shows how this method allows us to take into account off-perturbative instanton contributions, which are usually out of reach of most analytic methods in such systems.

Despite extensive theoretical and experimental efforts over many years, the nature of the phase transition in the random field Ising model is still a mystery. The main problem here is that in the vicinity of the phase transition point (similarly to the random bond ferromagnet) the mean-field stationary equations can have many solutions, and in such a situation the usual perturbative RG approach, at least in its traditional form, does not work. Although at the present state of knowledge in this field it is very difficult to say what the systematic approach to the problem could be, one of the possibilities is that the summation over numerous local minima states could be incorporated into the renormalization-group theory in a form of a particular replica symmetry breaking (RSB) structure of the parameters of the renormalized Hamiltonian. This idea is actively discussed in the literature [80]. At present there are many studies, both analytical and numerical [37, 79, 81], which indicate that owing to the existence of numerous local minima states, instead of the usual ferromagnetic phase transition point a special intermediate 'glassy' phase may set in, separating paramagnetic and ferromagnetic phases.

In the scope of directed polymers, in Chapter 12 we have considered only the most simple one-dimensional problems, for which the results for the wandering exponent are believed to be sufficiently well established. Nevertheless, even here, in the one-dimensional case with a locally correlated random potential, there are important problems that are still waiting for their solution. Presumably the most important one is that the Bethe ansatz solution considered in Chapter 12 cannot provide any physically sensible result even for the replica partition function (and for the corresponding average free energy) of the originally formulated problem with free boundary conditions. Even less clear is the situation in more sophisticated types of the directed polymer model. In particular, there are no solutions for the one-dimensional systems with the random potential characterized by (slow) long-distance algebraic decay of correlations, not to mention models in higher dimensions.

Thus, in conclusion, recent developments in the scope of statistical mechanics of disordered systems reveal a number of qualitatively new physical phenomena, and some of them go well beyond the traditional approaches of equilibrium statistical mechanics. Maybe we are close to a breakthrough to a new level of understanding of physics of disordered materials. In any case, this story is far from being finished.

Bibliography

[1] M. Mezard, G. Parisi and M. Virasoro, *Spin-Glass Theory and Beyond* (World Scientific, 1987).

[2] Vik. S. Dotsenko, 'Physics of Spin-Glass State', *Physics-Uspekhi*, **36**(6), 455 (1993).

[3] R. Rammal, G. Toulouse and M. A. Virasoro, *Rev. Mod. Phys.* **58**, 765 (1986).

[4] K. Binder and A.P. Young, 'Spin Glasses: Experimental Facts, Theoretical Concepts and Open Questions', *Rev. Mod. Phys.* **58**, 801 (1986); D. Chowdhury *Spin Glasses and Other Frustrated Systems*, (World Scientific, 1986); K.H. Fisher and J. Hertz, *Spin Glasses* (Cambridge University Press, 1991); Vik. S. Dotsenko, M.V. Feigelman and L.B. Ioffe, *Spin Glasses and Related Problems, Soviet Scientific Reviews*, Vol. 15 (Harwood, 1990).

[5] *Spin Glasses and Random Fields*, (*Series on Directions in Condensed Matter Physics*, Vol. 12) ed. A.P. Young (World Scientific, 1998).

[6] C.M. Newman and D.L. Stein, *Phys. Rev.* **B46**, 973 (1992); *Phys. Rev. Lett.* **76**, 515 (1996); *Phys. Rev. Lett.* **76**, 4821 (1996); *Phys. Rev.* **E57**, 1356 (1998); F. Guerra, *Int. J. Mod. Phys.* **B10**, 1675 (1996); E. Marinari, G. Parisi and J.J. Ruiz-Lorenzo, in *Spin Glasses and Random Fields* (*Series on Directions in Condensed Matter Physics*, Vol. 12) ed. A.P. Young (World Scientific, 1998); C. De Dominicis, T. Temesvári and I. Kondor, *J. Phys IV* (*France*), **8**, 13 (1998); C. De Dominicis, I. Kondor and T. Temesvári, in *Spin Glasses and Random Fields* (*Series on Directions in Condensed Matter Physics*, Vol. 12) ed. A.P. Young (World Scientific, 1998); E. Marinari, G. Parisi, F. Ricci-Tersenghi, J.J. Ruiz-Lorenzo and F. Zuliani, Preprint cond-mat/9906076.

[7] W.L. McMillan, *J.Phys.* **C17**, 3179 (1984); A.J. Bray and M.A. Moore, in *Heidelberg Colloquium on Glassy Dynamics* (Lecture Notes in Physics, Vol. 275) eds. J.L. van Hemmen and I. Morgenstern (Springer-Verlag, Heidelberg, 1987); A.J. Bray and M.A. Moore, *Phys. Rev. Lett.* **58**, 57 (1987); D.S. Fisher and D.A. Huse, *Phys. Rev.* **B38**, 373 (1988); *Phys. Rev.* **B38**, 386 (1988).

[8] L.F. Cugliandolo and J.Kurchan, *Phys. Rev. Lett.* **71**, 173 (1993); *J. Phys.* **A27**, 5749 (1994); *Philosophical Magazine*, **71**, 501 (1995); J.-P. Bouchaud, L.F. Cugliandolo, J. Kurchan and M. Mézard, in *Spin Glasses and Random Fields* (*Series on Directions in Condensed Matter Physics*, Vol. 12) ed. A.P. Young (World Scientific, 1998).

[9] M. Lederman *et al.*, *Phys. Rev.* **B44**, 7403 (1991); E. Vincent *et al.*, 'Slow Dynamics in Spin Glasses and Other Complex systems', in *Recent Progress in*

Random Magnets, ed D.H. Ryan (World Scientific, 1992); J. Hammann *et al.*, 'Barrier Heights Versus Temperature in Spin Glasses', *J.M.M.M.* 104–107, 1617 (1992); F. Lefloch *et al.*, 'Can Aging Phenomena Discriminate Between the Hierarchical and the Droplet model in Spin Glasses?', *Europhys. Lett.* **18**, 647 (1992).

[10] G. Parisi, *Statistical Field Theory* (Addison-Wesley, 1988); A.Z. Patashinskii and V.L. Pokrovskii, *Fluctuation Theory of Phase Transitions* (Pergamon Press, 1979); K.G. Wilson and J. Kogut, 'The Renormalization Group and the ϵ-expansion', *Physics Reports*, **12C** (2), 75 (1974).

[11] T.D. Schultz, D.C. Mattis and E.H. Lieb, *Rev. Mod. Phys.* **36**, 856 (1964)

[12] F.A. Berezin, *The Method of Second Quantization* (New York, Academic Press, 1966).

[13] C.A. Hurst and H.S. Green, *J. Chem. Phys.* **33**, 1059 (1960).

[14] F.A. Berezin, *Usp. Math. Nauk.* **24**, 3 (1969); V.N. Popov, *Functional Integrals in Quantum Field Theory and Statistical Theory* (Moscow, Atomizdat, 1976); E. Fradkin, M. Srednicki and L. Susskind, *Phys. Rev.* **D21**, 2885 (1980).

[15] T. Nattermann and J. Villain, *Phase Transitions* **11**, 5 (1988); T. Nattermann and P. Rujan, *Int. J. Mod. Phys.* **B3**, 1597 (1989).

[16] T. Halpin-Healy and Y.-C. Zhang, *Physics Reports* **254**, 215 (1995); M. Kardar, Lectures given at the 1994 Les Houches summer school on '*Fluctuating Geometries in Statistical Mechanics and Field Theory*', eds. F. David, P. Ginsparg and J. Zinn-Justin (North Holland, 1996).

[17] G. Toulouse, *Commun. Phys.* **2**, 115 (1977).

[18] S.F. Edwards and P.W. Anderson, *J. Phys.* **F5**, 965 (1975).

[19] D. Sherrington and S. Kirkpatrick, *Phys. Rev. Lett.* **35**, 1972 (1975).

[20] C. de Dominicis and I. Kondor, *Phys. Rev.* **B27** 606 (1983).

[21] J.R.L. de Almeida and D.J. Thouless, *J. Phys.* **A11**, 983 (1978).

[22] G. Parisi, *J. Phys.* **A13**, L115 (1980).

[23] M. Mezard *et al.*, *J. Physique* **45**, 843 (1984).

[24] M. Mezard and M.A. Virasoro, *J. Physique* **46**, 1293 (1985).

[25] Vik. S. Dotsenko, *J. Phys.* **C18**, 6023 (1985).

[26] M. Alba *et al.*, *J. Phys.* **C15**, 5441 (1982); E. Vincent and J. Hammann, *J. Phys.* **C20**, 2659 (1987).

[27] M. Alba *et al.*, *Europhys. Lett.* **2**, 45 (1986).

[28] L. Lundgren *et al.*, *Phys. Rev. Lett.* **51**, 911 (1983).

[29] R.W. Penney, T. Coolen and D. Sherrington, *J. Phys.* **A26**, 3681 (1993).

[30] V.S. Dotsenko, S. Franz and M. Mezard, *J. Phys.* **A27**, 2351 (1994).

[31] A.I. Larkin and D.E. Khmelnitskii, *JETP* **59**, 2087 (1969); A. Aharony, *Phys. Rev.* **B13**, 2092 (1976).

[32] M.B. McCoy *Phys. Rev.* **B2**, 2795 (1970); B.Ya. Balagurov and V.G. Vaks, *ZhETF (Soviet Phys. JETP)* **65**, 1600 (1973).

[33] A.B. Harris, *J. Phys.* **C7**, 1671 (1974).

[34] C. Domb, *J. Phys.* **C5**, 1399 (1972); P.G. Watson, *J. Phys.* **C3**, L25 (1970).

[35] A.B. Harris and T.C. Lubensky, *Phys. Rev. Lett.* **33**, 1540 (1974); D.E. Khmelnitskii, *ZhETF (Soviet Phys. JETP)* **68**, 1960 (1975); G. Grinstein and A. Luther, *Phys. Rev.* **B13**, 1329 (1976).

[36] M. Mezard and G. Parisi, *J. Phys. I* **1**, 809 (1991).

[37] M. Mezard and A.P. Young, *Europhys. Lett.* **18**, 653 (1992); M. Mezard and R. Monasson, *Phys. Rev.* **B50**, 7199 (1994).

[38] S. Korshunov, *Phys. Rev.* **B48**, 3969 (1993).

[39] P. Le Doussal and T. Giamarchi, *Phys. Rev. Lett.* **74**, 606 (1995).
[40] Vik. S. Dotsenko, B. Harris, D. Sherrington and R. Stinchcombe, *J. Phys. A: Math. Gen.* **28**, 3093 (1995).
[41] Vik. S. Dotsenko and D.E. Feldman, *J. Phys. A: Math. Gen.* **28**, 5183 (1995).
[42] A.M. Polyakov, *Gauge Fields and Strings* (Harwood Academic, 1987).
[43] D. Gross, I. Kanter and H. Sompolinsky, *Phys. Rev. Lett.* **55**, 304 (1985).
[44] R. Griffiths, *Phys. Rev. Lett.* **23**, 17 (1969).
[45] J.L. Cardy and A.J. McKane, *Nucl. Phys.* B **257** [FS14] 383 (1985); A.J. Bray, *Phys. Rev. Lett.* **59**, 586 (1987); A.J. Bray and D. Huifang, *Phys. Rev.* **B40**, 6980 (1989).
[46] A.T. Ogielski, *Phys. Rev.* **B32**, 7384 (1985); J.J. Ruiz-Lorenzo, *J. Phys. A: Math. Gen.* **30**, 485 (1997).
[47] L. Onsager, *Phys. Rev.* **65**, 177 (1944).
[48] Vik. S. Dotsenko and Vl. S. Dotsenko, *Adv. Phys.* **32**, 129 (1983); Vik. S. Dotsenko and Vl. S. Dotsenko, *JETP Lett.* **33**, 37 (1981); Vik. S. Dotsenko and Vl. S. Dotsenko, *J. Phys.* **C15**, 495 (1982).
[49] S. Sherman, *J. Math. Phys.* **1**, 202 (1960); N.V. Vdovichenko, *ZhETF* **47**, 715 (1964); *Ibid.* **48**, 526 (1965).
[50] L.D. Landau and E.M. Lifshits, *Statistical Physics*, 3rd edition (Oxford, Pergamon Press, 1980).
[51] N.B. Shalaev, *Sov. Phys. Solid State*, **26**, 1811, (1983); R. Shankar, *Phys. Rev. Lett.* **58**, 2466 (1984); A.W.W. Ludwig, *Phys. Rev. Lett.* **61**, 2388 (1988); R. Shankar, *Phys. Rev. Lett.* **61**, 2390 (1988).
[52] D.E. Feldman, A.V. Izyumov and Vik. S. Dotsenko, *J. Phys.* **A29**, 4331 (1996).
[53] Vik. S. Dotsenko, Vl. S. Dotsenko, M. Picco and P. Pujol, *Europhys. Lett.* **32**, 425 (1995).
[54] A.L. Talapov, V.B. Andreichenko, Vl. S. Dotsenko, W. Selke and L.N. Shchur, *JETP Lett.* **51**, 182 (1990) (*Pis'ma v Zh. Esp. Teor. Fiz.* **51**, 161 (1990)); V.B. Andreichenko, Vl. S. Dotsenko, L.N. Shchur and A.L. Talapov, *Int. J. Mod. Phys.* **C2**, 805 (1991); A.L. Talapov, V.B. Andreichenko, Vl. S. Dotsenko and L.N. Shchur, *Int. J. Mod. Phys.* **C4**, 787 (1993).
[55] V.B. Andreichenko, Vl. S. Dotsenko, W. Selke and J.-S. Wang, *Nuclear Phys.* **B344**, 531 (1990); J.-S. Wang, W. Selke, Vl. S. Dotsenko and V.B. Andreichenko, *Europhys. Lett.* **11**, 301 (1990).
[56] J.-S. Wang, W. Selke, Vl.S. Dotsenko, and V.B. Andreichenko, *Physica* **A164**, 221 (1990).
[57] A.L. Talapov and L.N. Shchur, 'Critical correlation function for the 2D random bonds Ising model', *Europhys. Lett.* **27**, 193 (1994); A.L. Talapov and L.N. Shchur, 'Critical region of the random bond Ising model', *J. Phys.* **C6**, 8295 (1994).
[58] R. Fish, *J. Stat. Phys.* **18**, 111 (1978); B. Derrida, B.W. Southeren and D. Stauffer, *J. de Physique (France)*, **48**, 335 (1987).
[59] R.H. Swendsen and J.-S. Wang, *Phys. Rev. Lett.* **58**, 86 (1987).
[60] A.E. Ferdinand and M.E. Fisher, *Phys. Rev.* **185**, 832 (1969).
[61] C. Jayaprakash, E.J. Riedel and M. Wortis, *Phys. Rev.* **B18**, 2244 (1978); J.M. Jeomans and R.B. Stinchcombe, *J. Phys.* **C12**, 347 (1979); W.Y. Cing and D.L. Huber, *Phys. Rev.* **B13**, 2962 (1976); C. Tsallis and S.V.F. Levy, *J. Phys.* **C13**, 465 (1980).
[62] R.B. Stinchcombe, in *Phase Transitions and Critical Phenomena* **7**, eds. C. Domb and J.L. Lebowitz (Academic Press, 1983).

[63] K. Binder and A.P. Young, 'Spin glasses: experimental facts, theoretical concepts and open questions', *Rev. Mod. Phys.* **58**, 801 (1986).
[64] W.L. McMillan, *Phys. Rev.* **B28**, 5216 (1983); W.L. McMillan, *Phys. Rev.* **B29**, 4026 (1984); W.L. McMillan, *Phys. Rev.* **B30**, 476 (1984).
[65] H. Nishimori, *Prog. Theor. Phys.* **66**, 1169 (1981); H. Nishimori, *Prog. Theor. Phys.* **76**, 305 (1986); Y. Ozeki and H.Nishimori, *J. Phys. Soc. Japan*, **56**, 3265 (1987).
[66] P. le Doussal and A.B. Harris, *Phys. Rev.* **B40**, 9249 (1989); Y. Ozeki and H. Nishimori, *J. Phys. Soc. Japan*, **56**, 1568 (1987); Y. Ueno and Y. Ozeki, *J. Phys. Soc. Japan*, **64**, 227 (1991); R.R.P. Singh, *Phys. Rev. Lett.* **67**, 899 (1991).
[67] D.P. Belanger, *Phase Transitions* **11**, 53 (1988).
[68] J. Villain, *J. de Physique* **43**, 808 (1982).
[69] P.G. de Gennes, *J. Phys. Chem.* **88**, 6449 (1984).
[70] J.F. Fernandez, *Europhys. Lett.* **5**, 129 (1985).
[71] Y. Imry and S.-K. Ma, *Phys. Rev. Lett.* **35**, 1399 (1975).
[72] J. Imbrie, *Phys. Rev. Lett.* **53**, 1747 (1984).
[73] A.P. Young, *J. Phys.* **C10** L257 (1977);
A. Aharony, Y. Imry and S.-K. Ma, *Phys. Rev. Lett.* **37**, 1364 (1976).
[74] G. Parisi and N. Sourlas, *Phys. Rev. Lett.*, **43**, 774 (1979); G. Parisi, *Quantum Field Theory and Quantum Statistics* (Bristol, Adam Hilger, 1987).
[75] G. Parisi and Vik. S. Dotsenko, *J. Phys. A: Math. Gen.* **25**, 3143 (1992).
[76] Vik. S. Dotsenko, *J. Phys. A: Math. Gen.* **27**, 3397 (1994).
[77] V. Villain, *Phys. Rev. Lett.* **52**, 1543 (1984); G. Grinstein and J. Fernandez, *Phys. Rev.* **29**, 6389 (1984); U. Nowak and K.D. Usadel, *Phys. Rev.* **B43**, 851 (1991).
[78] G. Parisi, *Proceedings of Les Houches 1982, Session XXXIX*, eds. J.B. Zuber and R. Stora (North Holland, Amsterdam, 1984); J. Villain, *J. Physique* **46**, 1843 (1985).
[79] M. Guagnelli, E. Marinari and G. Parisi, *J. Phys.* **A26**, 5675 (1993).
[80] H. Yoshizawa and D. Belanger, *Phys. Rev.* **B30**, 5220 (1984); Y. Shapir, *J. Phys.* **C17**, L809 (1984); C. Ro, G. Grest, C. Soukoulist and K. Levin, *Phys. Rev.* **B31**, 1682 (1985); J.R.L. de Almeida and R. Bruisma, *Phys. Rev.* **B35**, 7267 (1987).
[81] C. De Dominicis, H. Orland and T. Temesvari, *J. Physique I* **5**, 987 (1996).
[82] M. Kardar and Y.-C. Zhang, *Phys. Rev. Lett.* **58**, 2087 (1987).
[83] M. Mezard and G. Parisi *J. Phys. I France* **1**, 809 (1991).
[84] M. Kardar, G. Parisi and Y.C. Zhang, *Phys. Rev. Lett.* **56**, 889 (1986); E. Medina, T. Hwa, M. Kardar and Y.-C. Zhang, *Phys. Rev.* **A39**, 3053 (1989).
[85] D.A. Huse, C.L. Henley and D.S. Fisher, *Phys. Rev. Lett.* **55**, 2924 (1985).
[86] M. Kardar, *Nucl. Phys.* **B290**, 582 (1987).
[87] S. Korshunov and Vik. Dotsenko, *J. Phys. A: Math. Gen.* **31**, 2591 (1998).
[88] Vik. S. Dotsenko and M. Mezard, *J.Phys. A: Math. Gen.* **30**, 3363 (1997).
[89] J. Villain, B. Semeria, F. Lanon and L. Billard, *J. Phys.* **C16**, 2588 (1983).
[90] J. Villain, *J. Phys.* **A21**, L1099 (1988).
[91] A. Engel, *J. Physique Lett.* **46**, L409 (1985).
[92] J. Villain and B. Séméria *J. Physique Lett.* **44**, L889 (1983).
[93] M. Mézard and G. Parisi, *J. Physique I* **2**, 2231 (1992).
[94] A. Engel, *Replica Symmetry Breaking in One Dimension* (Gottingen University preprint, 1993).
[95] F. Cesi *et al.*, *Comm. Math. Phys.* **188**, 135 (1997); F. Cesi *et al.*, *Comm. Math. Phys.* **189**, 323 (1997).

[96] Vik. S. Dotsenko, *J. Phys. A: Math. Gen.* **32**, 2949 (1999).
[97] J. Zinn-Justin, *Quantum Field Theory and Critical Phenomena* (Clarendon Press, Oxford 1989, third edn. 1996).
[98] L.N. Lipatov, *JETP Lett.* **24**, 157 (1976); *Sov. Phys. JETP* **44**, 1055 (1976); *JETP Lett.* **25**, 104 (1977); *Sov. Phys. JETP* **45**, 216 (1977).

Index